Beetles in conservation

Beetles in conservation

T.R. New

A John Wiley & Sons, Ltd., Publication

Registered office: John Wiley & Sons Ltd, The Atrium, Southern Gate, Chichester, West Sussex, PO19 8SQ, UK

Editorial offices: 9600 Garsington Road, Oxford, OX4 2DQ, UK
The Atrium, Southern Gate, Chichester, West Sussex, PO19 8SQ, UK
111 River Street, Hoboken, NJ 07030-5774, USA

For details of our global editorial offices, for customer services and for information about how to apply for permission to reuse the copyright material in this book please see our website at www.wiley.com/wiley-blackwell

Library of Congress Cataloguing-in-Publication Data

New, T. R.
    Beetles in conservation / T.R. New.
        p. cm.
    Includes bibliographical references and index.
    ISBN 978-1-4443-3259-9 (alk. paper)   1. Beetles–Conservation.   2. Beetles–Ecology.
I. Title.
    QL573.N49 2010
    595.76–dc22
                                                                                    2009038743
ISBN: 978-1-4443-3259-9

A catalogue record for this book is available from the British Library.

Set in 10.5/12pt Classical Garamond by Graphicraft Limited, Hong Kong
Printed and bound in Singapore by Fabulous Printers Pte Ltd

1   2010

Contents

Preface

Some years ago I discussed my tentative ideas for a book on beetle conservation with several experienced coleopterists. Eminent coleopterist number 1 gave his initial reaction that 'no one in his right mind would try it'. This encouragement was echoed by eminent coleopterist number 2 in suggesting that the subject was just too vast for anyone to tackle properly, so that any author's credibility might be placed seriously at stake. As Zimmerman (1994) put it, at the commencement to volume I of *Australian Weevils*, 'Those who know most about this subject know best how little they know'. As a non-coleopterist, I recognize very clearly the dangers of a little knowledge, but during several decades of interest in insect conservation I have gained some general appreciation of the importance of beetles in the natural world, and of conserving them and the systems they help to sustain. I have become aware of some of the many ways in which knowledge of beetle ecology and population management has contributed significantly to wider conservation progress and interest. However, with a lifetime passed among the 'little folks' of entomology, I do not aspire to join the men of Oliver Wendell Holmes' essay[1], although I recall thinking the comparison a little unfair when I came across it first (in Walsh & Dibb 1954) as a young teenager. However, from this vantage point, and *mens sana* or no, such a synthesis or overview appears worthwhile, if for no other reasons than to challenge conservation-oriented coleopterists to refine, correct and expand on the perspective presented here, and to acknowledge and emphasize the important place of Coleoptera in the wider scenarios of insect conservation as this science matures.

Interest in beetles extends for well over two centuries, and has spawned a vast and daunting literature on these insects, collectively addressing many aspects of their identification and biology. In recent years, printed material has

---

[1] 'These Lepidoptera are for children to play with, pretty to look at, so some think. Give me the Coleoptera, and the kings of the Coleoptera are the beetles! Lepidoptera and Neuroptera for little folks; Coleoptera for men, sir!' Oliver Wendell Holmes (1872) *The Poet at the Breakfast Table*. Walter Scott: London.

been augmented enormously by web-based information and comment, so that entering the name of any major beetle group into a recognized search engine is likely to generate hundreds to hundreds of thousands of responses. Refinement of search terms yields much of conservation interest, not least through facilitating access to unpublished reports and other forms of 'grey' literature. Devotees of any particular beetle group are likely to be served by specialist newsletters and, in some cases, journals as important foci for their interests. A number of more generalized Coleoptera journals occur, some as organs for society or specialist groups. The literature on pest Coleoptera is also enormous, with their management commonly a major economic need. The collector interests so important in fostering conservation awareness for insects are increasingly catered for by series of handbooks (such as Cooter & Barclay 2006, for Britain) or identification guides to more popular families, particularly of larger beetles. The increasing availability and use of colour to illustrate many of these is evident in many publishers' and dealers' catalogues of modern books, with a correspondingly deep pocket needed to purchase many of these volumes. Many, indeed, seem priced to ensure that their distribution is restricted to wealthier institutional libraries but, significantly, also that they are unlikely to reach important potential user groups in less-developed parts of the world. However, whilst books on beetles abound, few focus specifically on beetle conservation or their importance in natural ecosystems. Many accounts allude to scarcity or rarity of particular species or more generally mention threats to beetles, but a very high proportion of major works expressing concern about the well-being of beetles do this simply as an apparent afterthought rather than a primary message or purpose. Likewise, the journal literature on beetle conservation is very widely dispersed. A recent special issue of *Journal of Insect Conservation* (2007) focused on beetles and conservation.

Nevertheless, over the last few decades in particular, a considerable variety of beetle species have appeared on conservation agendas, for a wide array of reasons. Some are listed formally among endangered or threatened species compilations, with such concerns arising from habitat changes, implications of over-exploitation by collectors or traders, and numerous other manifestations of human activity that render them increasingly vulnerable. Understanding conservation need draws on the massive amount of knowledge on how beetles work, much of this emanating from studies on pest species or beneficial species, such as predators of crop pests. In this book, I have tried to bring together some of the information on relationships between beetles and their environments, and to exemplify the roles of beetles as individual targets for conservation, and as broad ecological tools of massive value in helping to assess environmental quality and change.

I cannot claim to have read more than a small fraction of the voluminous literature on beetles, to which I allude above, but have tried to digest a reasonable selection of the papers and essays (published up to late 2008) in which appraisal and management for conservation has been discussed. The first chapter is a broad introduction to beetle conservation, diversity and ecology, illustrating many of the major advantages and disadvantages in employing such a vast and varied group, and how beetles may be incorporated constructively into conservation theory and practice. Chapter 2 outlines some of the main approaches, and how beetles may be studied and appraised for conservation. Threat evaluation is a paramount theme, whether to individual species, assemblages or wider habitats, and the most universal of these devolves on changes to places and resources, broadly 'habitats', and understanding how those changes affect the occupants or users. Studies of the responses of beetles to environmental changes have contributed substantially to this understanding, in ways that can be extrapolated easily to concerns for many other invertebrates. Background to the suite of themes involved is included in Chapter 3. The later chapters are shorter. Chapters 4–6 exemplify the variety of other threats, one or more of which may become important, predictably or unexpectedly, and perhaps exacerbate effects of habitat change on other species. One major threat with severe implications, climate change, is both inevitable and at this stage largely imponderable in detail, but emphasizes the need for a long-term view in insect conservation programmes. The tactics available for conservation management include establishment of new populations, through a variety of *ex situ* techniques which are discussed in Chapter 7. Parameters of habitat and resource management are numerous but, as indicated in Chapter 8, there may be a very fine line between benefits from constructive manipulations and enhancing threats through well-intentioned changes but ignorance of the wider ramifications of these. Chapter 9 includes further general background to several of the families of beetles that have contributed most constructively to advancing knowledge of beetle conservation and, in the final chapter (Chapter 10) I discuss some possible ways in which the future of this predominant insect group may be rendered more secure.

T.R. New
Department of Zoology
La Trobe University
Melbourne

Acknowledgements

The following are thanked for permission to use or modify previously published material in this book: CAB International, Wallingford, UK; Cambridge University Press, Cambridge; Ecological Society of America; Entomological Society of America; Elsevier Science, Oxford; Finnish Zoological and Botanical Publishing Board, Helsinki; National Academy of Sciences, USA; Oxford University Press, Oxford; Princeton University Press, Princeton, New Jersey; Springer Science and Business Media, Dordrecht; Surrey Beatty and Sons Pty Ltd, Baulkham Hills, New South Wales, Australia; Wiley-Blackwell, Oxford. Every effort has been made to obtain permission for such use. The publisher apologises for any inadvertent errors or omissions and would welcome news of any corrections that should be incorporated in future reprints or editions of this book.

This book was initiated through Ward Cooper at Wiley-Blackwell. I greatly appreciate his continuing support, together with the help given by Delia Sandford, Camille Poire, Kathy Auger and Kelvin Matthews during early production stages. Copy-editing by Jo Phillips is gratefully acknowledged.

The cover illustration is from an original embroidery sewn by Zanela Buthelezi Matsetsa, whom I thank for allowing me to use it here. Thanks are due also to Kathy Ducasse, Brigid Turner and Shirley Hanrahan for their help in facilitating this.

1

Introduction

Beetles and conservation

Beetles, the members of the insect order Coleoptera, are widely believed to be the most species-rich animal group that exists on Earth and, perhaps, that has ever shared our world. Comprising around one-quarter of all animal species, their richness and ubiquity led Evans and Bellamy (1996) to comment 'We live in the Age of Beetles'. That Age is a long one. Modern beetles, the outcomes of some 250 million years of evolution since the earliest beetle fossils found in the Permian period, are difficult to ignore. Collectively, they are immensely diverse in their lifestyles and ecology, and intrude on human consciousness and well-being in many ways. Our perceptions of beetles cover a huge range of human experience: as creatures of cultural importance and symbolism (such as scarabs in ancient Egypt), as objects of desire and fascination to naturalists and collectors, and as severe pests and our major competitors for crops and other commodities. Their countering positive values are as predators and biological control agents of other pests, as valued tools in environmental assessments of terrestrial and freshwater ecosystems, as study tools to help elucidate ecological functions, and as key components aiding the working and sustainability of the natural world. Much of the above emphasizes their ecological variety, with more than half the order in some way phytophagous. Plant-feeding apparently arose early in beetle evolution, with the ancestors of the current major radiations of angiosperm-feeding beetles (namely the weevils, Curculionoidea, and leaf beetles, Chrysomeloidea) existing in the Triassic, some 230 million years ago, and arising most likely from conifer- and cycad-feeding lineages (Farrell 1998). The massive radiations of beetles in the Cretaceous period paralleled the development and spread of flowering plants, and over about 100 million years

Beetles in Conservation, 1st edition. By T.R. New. Published 2010 by Blackwell Publishing.

they developed a broad array of feeding guilds and occupied virtually all available biotopes (see Erwin & Geraci 2009). However, numerous beetles are predators, fungus-feeders or other, and the collective variety of feeding habits encompasses exploitation of the variety of accessible foodstuffs in natural and anthropogenic environments.

Thus, unlike butterflies, whose wide general appeal renders them invaluable ambassadors in promoting the values of insect conservation and which contain few pest or otherwise damaging or undesirable species, the public image of beetles is decidedly mixed and more complex. Indeed, the same species may be regarded as a pest or conservation target in different places and its status change over time, so that local and wider perceptions of its role (and sympathy for its conservation) may differ substantially across its range. One example is the oak pinhole borer beetle *Platypus cylindrus* (Platypodidae). This species is a serious forestry pest in parts of continental Europe, but was listed as rare and associated with veteran oaks in Britain (Shirt 1987). Since then, and following severe gales in Britain in 1987, it has become a serious pest of stressed and dead oak trees. Yet, whatever the practical perceptions of particular beetles may be, the group's richness and diversity renders them of immense importance in understanding the natural world and leads to human interests encompassing the extremes of conservation on the one hand to suppression or eradication on the other. The latter attitude may broaden from confirmed pest species to others, just in case they cause damage. And the various components of 'beetlephilia' (a term advanced by Evans & Bellamy 1996, drawing on E.O. Wilson's famous 'biophilia') ensure continued attention to some groups of beetles by hobbyists and others not concerned directly with either their welfare or slaughter. Perhaps tiger beetles (treated variously as Carabidae: Cicindelinae, or the full family Cicindelidae), many of which are brightly coloured and active by day, promote such attitudes particularly well, so that Pearson *et al.* (2006) introduced their book with a chapter entitled 'The magic of tiger beetles', and included comments such as 'Hundreds of otherwise normal people are passionate about an intriguing group of insects called tiger beetles' and 'tiger beetles elicit something more than a routine response to the necessities of employment'.

The appearance of many beetles can be dramatic, even bizarre; for example, the enlarged mandibles or pronotal horns of some Lucanidae or Scarabaeidae have long been objects of curiosity and appeal. Arrow's (1951) sentiment at the start of his book on these insects ('all who see . . . one of the great horned beetles for the first time cannot fail to experience feelings of astonishment') remains entirely suitable, whilst many smaller beetles may impress just as much by some feature of colour or morphological extravagance.

In short, such liking and sympathy is an important positive component of insect conservation. Perhaps particularly for beetles embedded in national or regional culture, public interest in conservation can be garnered readily. Thus, the Genji firefly (*Luciola cruciata*, Lampyridae) in Japanese traditional agricultural environments has always attracted exceptional public interest (Takeda *et al.* 2006) and is an important flagship species for conservation. The earliest literature records of these fireflies are reportedly in Japan's oldest collection of poetry in the late eighth century (Masayasu 2005), and their flight season (in early summer:

June, July) attracts tourists through events such as the annual Yokoyama Firefly Village festival that can promote interest in conservation. Larvae of *L. cruciata* are aquatic, and the species has suffered greatly from habitat loss and degradation through pollution, following likely over-collecting for sale in the past.

It may indeed be feasible to promote beetles (and many other invertebrates) responsibly for greater interest in ecotourism itineraries. Buprestidae and Scarabaeidae were listed among the sample invertebrate groups suitable for inclusion in such activities in South Africa (Huntly *et al.* 2005), the former as conspicuous spectacular-looking beetles often seen on flowers, and the latter for their dung-rolling activities. The Addo elephant dung beetle (see p. 74) has received particular attention, but the suggested approach for invertebrates has been to highlight particular features as part of an educational process. In this example, Buprestidae were promoted by contrasting the habits of pollen-feeding adults with wood-boring larvae, and dung beetles for their important ecological role in breaking down wastes and their elaborate behaviour (Huntly *et al.* 2005). These families were selected as among those easily seen in one game reserve, with ease of observation an important aspect of promoting insects to visitors. Surveys of tourists in South Africa have suggested that many people will embrace chances to broaden their experience beyond the major current focus on large mammals (mainly the Big Five) and to learn about other taxa. An important requirement to facilitate this is to train tour guides more effectively, so they can comment on invertebrates as well as the larger animals. Conversely, epidemics of pest beetles may deter tourism. The massive outbreak of mountain pine beetle *Dendroctonus ponderosae* in western Canada (see p. 134) has the potential to affect both the ecology of some national parks and the experiences gained by visitors (McFarlane & Witson 2008). Likely consequences on visitors include lessening quality of scenery, hazards from dead and falling trees, and effects on local economics by reduced tourist numbers, with this effect extending well beyond park boundaries. However, the role of bark beetle epidemics in public perception can be more complex. Whilst they are massively damaging to commercial forests, in national parks the beetles may be regarded instead as natural regeneration agents, helping to sustain the area's ecosystems (Muller & Job 2009). Despite the reaction of visitors noted above for Canada, surveys in an affected park in Bavaria (Germany) showed that the infesting beetle (*Ips typographus*) was generally accepted by tourists. A prevailing opinion was that control measures should not be introduced in the park (Muller & Job 2009). Information provided about the beetle's function and importance countered initial negative attitudes, as an education process fostering wider appreciation.

However, and perhaps paradoxically, disliking economically damaging or other pest beetles may also be important in promoting interests in beetles, because it leads to accumulation of information with considerable relevance in conserving their close relatives or other species occurring in similar environments. Some pest beetles, in stored products, timber, or as pests of agricultural, orchard or forestry crops, are among the best-studied insects. Much of that knowledge, as well as that on beneficial species such as manipulable predatory ladybirds (Coccinellidae) used as biological control agents, and the techniques by which it is acquired and analysed merits careful appraisal by insect conservationists.

The juxtaposition between the applied entomology literature and the conservation ecology literature is perhaps nowhere closer than for Coleoptera, and the wealth of detail in the former can provide invaluable ideas and leads in management of beetle species and assemblages for conservation. The appraisal of Coccinellidae by Dixon (2000), for example, is a broad biological foundation of interest for studying any member of that family in a wider perspective.

As well as contributing enormously to evolutionary and ecological understanding in numerous different terrestrial and freshwater ecosystems, beetles are important considerations in practical conservation. Their value and their roles range from the high-profile focus on single notable species threatened with decline or loss to documentation of the vast regional assemblages reflecting dependence on restricted resources threatened by human activity. Regarded widely as an easily sampled taxon, beetles have become an important group in addressing many questions of wide conservation relevance, so contributing to disciplines such as landscape ecology, reserve design and placement, and restoration or rehabilitation of most terrestrial and freshwater biomes. The bulk of ecological studies on beetles have not been undertaken specifically to address conservation issues. However, the leads they give, the background information accumulated on more basic ecology and biological understanding of how beetles work, and the methods and analyses pursued collectively lay a very sound foundation for more conservation-focused endeavours. It is also increasingly common for the discussions in papers on beetle biology to allude to the conservation implications of the work presented. It is impossible to summarize the ecology of beetles briefly, but their ecological specializations and variety may be considered to lie along a continuum from strictly monophagous and highly specialized species (with obligately small niches and frequenting only one or two habitat types, sometimes highly circumscribed as local endemics) to extreme generalist species (with wide niche breadths and in a wide variety of habitats across a broad geographical range) (Dufrene & Legendre 1997). Any terrestrial or aquatic beetle assemblage in a restricted habitat will thus include the two major elements of (i) obligate specialist species restricted or largely restricted to it, and (ii) generalists that may either extend casually into that environment or, by actively selecting particular resources or other attributes, become more abundant there than elsewhere. Whereas conservation interest may gravitate predominantly towards the specialist species, which tend to become detectably threatened more easily than many generalists, the operating environment for such species includes the assemblage of which they are part, and with which they may interact. Slightly more broadly, it is useful to separate three major ecological categories that transcend any individual trophic category: (i) ubiquitous species, i.e. those that are geographically and ecologically wide-ranging; (ii) eurytopic species, i.e. those found in a variety of habitats but over a more restricted geographical range; and (iii) stenotopic species, i.e. those that are much more specific and found in one or few habitats, as specialists. The first group, ubiquitous species, are commonly also eurytopic, because they occur in a variety of habitats, so are then distinguished by the extent of their distribution.

Highly specialized ecological oddities abound among Coleoptera, so that many broad generalizations about their habits and biology are subject to increasing numbers of exceptions as biologically novel or unexpected traits are discovered. Some may augment conservation interest as informative evolutionary lineages, or by demonstrating unusual adaptations that render them resistant or vulnerable to environmental changes. One recently appraised oddity is the fairy shrimp hunting beetle (*Cicinis bruchi*), a highly unusual carabid (Erwin & Aschero 2004). Unlike most ground beetles, this species (formerly known from only two specimens and characterized, in an intriguing image, as 'the carabid equivalent of a crocodile') is aquatic and the nocturnal adults swim on the surface of alkaline water bodies, salt flats, in Argentina. The salt flats are very extensive, but Erwin and Aschero (2004) believe that the beetle's distribution there was restricted by that of 'tiled' soils providing refuges for the beetles during the day. The beetles feed solely on anostracan shrimps. At present, the extensive habitat area seems not to be threatened, but the future influence of global warming may pose a severe threat in this semi-desert environment.

Beetle extinctions and extirpations

In his magisterial *The Biology of Coleoptera*, Crowson (1981) commented (p. 650):

> The immense diversity of the recent beetle fauna of the earth is the product of extremely diversified ecosystems which have maintained considerable degrees of stability for very long periods of time. Both the diversity and the stability are now being destroyed, at an increasing rate, by human action. In the process large numbers of Coleoptera must inevitably become extinct . . .

and (p. 689) 'we fear . . . the early extinction of large and scientifically interesting parts of the present world fauna of Coleoptera'. Stability or stasis of many beetle species is inferred strongly from the record of Quaternary beetle fossils (or subfossils), which comprise fragments that resemble closely, and indeed are often identical to, the equivalent parts of modern beetles. Many of these, including elytra, pronota, heads and sclerotized male genitalia, enable fossils to be identified reliably as modern species. Many such fossils (which can occur in large numbers in sediments) retain structural and pigmented colours, together with setae and micro-ornamentation, and provide strong indication of the long periods for which some species have existed, notwithstanding the many gaps in the record. Crowson's sentiment has been echoed repeatedly, for example by Erwin (1997) in referring to the vulnerability of tropical forest beetles: 'current human activity and that of the immediate future will exterminate a large percentage of these species'. In general, fear of extinction of enormous numbers of species, including beetles, within the next few decades (Dunn 2005) is a vital rallying call for urgent conservation measures throughout the world. More specific contexts occur, with many general warnings of likely demise of particular beetle taxa. For example, surveys of the large montane bess beetles (Passalidae) roused the comment from Schuster *et al.* (2003, p. 302) that 'In general *Proculus*, as well as other montane species of passalids, is probably in danger of extinction throughout its range due to the elimination of most of the forest where it occurs'.

With few exceptions, however, specific knowledge of such extinctions does not exist, particularly over most of the tropics where beetle diversity can be extraordinarily high. Indeed, Mawdsley and Stork (1995) could enumerate only 10 recorded global species extinctions of beetles, all of them from isolated islands. Even in the Quaternary fossil record, beetle extinctions seem to be few, with stasis (accompanied by, sometimes dramatic, range changes; p. 135) more common, reflecting the sequences of glacial and interglacial periods over which vegetation may change through advance and retreat over several thousand kilometres. Coope (1995) and others have suggested that the perceived lack of extinctions may be partly due to this long-term series of changes, associated with a high degree of population mixing with resulting homogeneity of populations diminishing the chances of extinction by countering genetic impoverishment. Indeed, Ashworth (2001) could cite only two possible extinctions over that period, the dung beetles *Copris pristinus* and *Onthophagus everestae* from the La Brea asphalt deposits, but, following Miller (1997), suggested caution in declaring that even these Pleistocene species had really disappeared. The twin concepts of structural extinction and sampling extinction differentiated by Gaston and McArdle (1994) are important components of understanding, whereby the latter emphasizes the difficulties of evaluating small, elusive and perhaps cryptic taxa. This dichotomy was explored by Didham *et al*. (1998a,b) in their study of forest fragmentation effects on beetles in Amazonia, and in which they attempted to appraise the likelihood of extinction from changing abundance across sites of different sizes and conditions (see p. 96). Habitat fragment size was apparently a key predictor of extinction risk for some species. Some insect extinctions are indeed difficult to prove, and declarations of extinction may reflect periods where the species is not recorded, but often with little indication of the amount of targeted search effort for it. It is not particularly uncommon for beetles to be rediscovered decades after they were last seen. One such example is the New Zealand dytiscid water beetle *Rhantus plantaris*, described from a single specimen in 1882 and found again in 1986 (Balke *et al*. 2000). The site where it was then found was a small perennial pond with water diameter only about 5 m, so that its existence may still be regarded as rather tenuous, but at least in 1986 it was not extinct! Whilst McGuinness (2001) remarked that three beetle species had been reported to be extinct in New Zealand, his own inferences were more cautious. For the carabid *Mecodema punctellum*, not seen since 1931, McGuinness (2002) noted 'this species *may be* extinct' (my italics). However, sometimes repeated and specifically targeted surveys have not revealed the beetle sought. The large flightless ground weevil *Hybomorpus melanosomus*, endemic to Lord Howe Island, formerly occurred there under logs and in rotten wood. It is known from a few specimens in collections and has not been collected since the 19th century. Soon after, it was considered to be extremely rare or possibly extinct (Oliff 1889), even before the introduction of rats to Lord Howe Island early in the 20th century. Intensive invertebrate surveys over several recent decades have not yielded specimens, and the weevil is now listed as presumed extinct under New South Wales Government legislation.

Strong declines and more local disappearances of particular beetle species are documented more commonly and effectively for parts of the temperate regions, particularly in the UK (from where Hambler & Speight 1996 listed 12 species believed to have become extinct since 1900, and Hyman & Parsons 1992, 1994 noted a number of species that had not been seen for at least several decades, but did not categorize these as extinct in their rankings), parts of western continental Europe, and parts of North America. These declines and extirpations are the source of much modern conservation interest, with the species brought to attention in this way the usual candidates for conservation. Studies on islands have also led to documentation of many such extirpations: several Tenebrionidae have been lost from particular Iberian islands for example (Cartagena & Galante 2002). In contrast to full extinctions, local extirpations are frequent and many of the threats noted later for recent beetles have had influences well before people became concerned about them (Whitehouse 2006). The riparian beetle faunas in northern Europe include a number of Carabidae that have been lost through river regulation and canalization, changes to nearby vegetation and bank structure, and pollution, with some peculiar to areas with particular substrates such as stones or sand (Andersen & Hanssen 2005). As another example, the Californian tiger beetle *Cicindela tranquebarica joaquinensis* was historically found over much of the San Joaquin Valley, associated with alkaline habitats. However, most populations have been extirpated because of intensive agricultural development, such as cultivation and changes of water for irrigation supply, so that the specialized habitat has been largely lost. Only three populations, each on a patch of habitat less than 3 ha in area, were known to Knisley and Haines (2007). As another striking example, the historical distribution of the American burying beetle *Nicrophorus americanus* (p. 145) was formerly extensive across the eastern half of the USA, but has now been reduced to three small disjunct areas (Fig. 1.1) (Lomolino *et al.* 1995; Sikes & Raithel 2002). Collectively, most of the more reliably evaluated recent losses are within the areas for which beetle faunas have been described most completely, as a continuing legacy of collector interests spanning some 150–200 years and the progressive availability of series of identification guides and handbooks that render the fauna at least partially tractable to people taking up their study.

Beetle diversity

Elsewhere, our knowledge of modern beetle diversity and its distribution is highly uneven, although broad historical biogeographical patterns within the order can be traced with the aid of the substantial fossil legacy (Coope 1995), so that many of the better-studied regional faunas can be defined and, in many instances, alien species recognized reliably. Thus, Australian beetles are often recognizable as regional endemics, and their presence elsewhere in the world (be they pests, beneficial species or with more neutral impacts) definable. Conversely, beetles from elsewhere are commonly detectable in Australia. Very commonly, knowledge from studies on alien beetles, undertaken to clarify their impacts or management to suppress or foster them, comes to exceed that available from within their natural range, and may have direct applications in conservation. Likewise, searches for beneficial insects, such as biological control agents, can

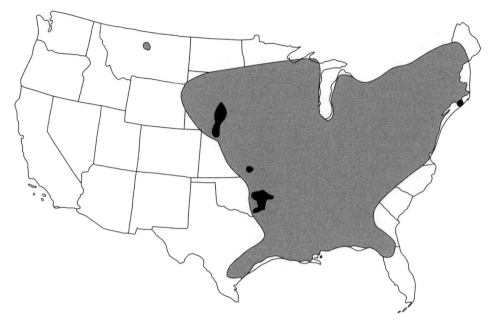

Fig. 1.1 The historical (shaded) and current (black) distribution range of the American burying beetle *Nicrophorus americanus* in North America. (After Lomolino *et al*. 1995 with permission.)

stimulate detailed investigation of possible source faunas: for example, Koch *et al.* (2000) noted the 'intensive, wide-ranging dung beetle collection programme' and resulting comprehensive reference collection of specimens from Australian exploration for South African dung beetles suitable for importation to Australia. However, in broad terms for much of the rest of the world, beetles are much less effectively documented than in Europe or North America; even though a strong systematic framework exists, many families have not been documented completely and a high proportion of species are unnamed and undiagnosed. We have little realistic idea about how many species of beetles occur on Earth; certainly several hundred thousand species have been described, but estimates of their total richness extend to several million species, and these figures continue to be debated. A tally of 358,000 described species (Bouchard *et al.* 2009) includes several earlier authoritative estimates, with the six largest families enumerated being Curculionidae (60,000 described species), Staphylinidae (47,744 species), Chrysomelidae (36,350 species), Carabidae (30,000 species), Scarabaeidae (27,800 species) and Cerambycidae (20,000 species). As Grove and Stork (2000, p. 735) commented 'In reality, and despite the best efforts of a number of researchers, we are still little nearer to determining the true extent of beetle diversity'. And, from Zimmerman (1994), 'Estimates of the numbers of described weevils are as variable as the opinions of those making the estimates'. However, several families are very large, whilst many others are small. The bulk of described beetles belong to only about eight families including those listed above, namely Carabidae, Staphylinidae, Scarabaeidae, Buprestidae, Tenebrionidae, Cerambycidae, Chrysomelidae and Curculionidae. An estimate by Gaston (1991) suggested that these together contained about two-thirds of beetles decribed at that time, and that high proportions of undescribed species would also be referable to these groups. Simply studying these families, whose members range over many trophic guilds and habitats, would alone provide a very strong practical framework for conservation.

However, considerably greater variety occurs. Thus, for Australia, around 23,000 beetle species have been described (in more than 120 of the global total of 166 families listed by Lawrence & Newton 1995 and, as an aside, a scarab beetle *Haploscapanes barbarossa* was the first formally named Australian endemic animal), but predictions of 80,000–100,000 species (or even more) have been made (Yeates *et al.* 2003). Such uncertainties are common, and render many aspects of evaluation based on fundamental documentation of biodiversity difficult and sometimes unconvincing, little more than 'guesstimates'. Within such large faunas, with their characteristic high levels of endemism, many beetles are ecologically specialized, and many are scarce or highly localized. Background knowledge and sampling effort is, in most cases, simply inadequate to detect their loss or continued presence in very low numbers, even if they are recognized as distinct entities. However, valuable information on assemblages can accrue relatively easily, because a high proportion of beetles can be allocated with reasonable confidence to feeding habit or guild, from knowledge of related taxa elsewhere. As an example from the tropics, in the poorly described beetle fauna of Sarawak, Malaysia, Chung *et al.* (2000) could confidently assess over 40% of the more than 1700 species they accumulated as predators, as well as determining that more than 15% were saprophages and fungivores and 10–13% herbivores, so that assemblage changes based on changing frequency of such guilds could be estimated and compared. Nevertheless, a high proportion of beetle species diagnosed or named from most of the tropical regions are known from few individuals, many of them from single specimens, and the biology of most of these is simply unknown, and can be inferred only in general terms by comparison with any better-known related taxa. Many beetles are known only from an inadequately documented, sometimes old, type specimen or description. The reality of studying beetles, and assessing their relevance to ecological sustainability and needs for conservation, involves acceptance of this vast uncertainty and learning how to treat it responsibly.

However, despite the taxonomic uncertainties which ensure that only low proportions of species in most general surveys of tropical beetles may be identifiable to species level (see below), studies of assemblage composition and the changes associated with disturbance or changing patterns of land use, made either along gradients (see p. 102) or more patchily, have commonly utilized beetles as signals or indicators of habitat condition. Many analyses of beetle assemblages have focused on particular families, so that differences or changes in richness and composition are correlated, sometimes tentatively, with habitat characteristics.

In any part of the world, and with information encompassing most continental and island areas, many beetles are regional or much more localized endemics, and most of the species signalled individually for conservation concern have highly restricted distributions. Additionally, Hammond (1994) recognized a category of near-endemics, illustrated by a number of intertidal/coastal beetles in Britain, but which are scarce or very restricted elsewhere in Europe. For example, some Staphylinidae in this category are known from very few sites outside Britain, and then only from Europe-facing Channel coasts or similar restricted ranges. For most parts of the world, patterns such as this cannot be defined with confidence, but there is no doubt that levels of narrow-range endemism among beetles can be very high. Two Australian examples illustrate the scenarios likely to be paralleled widely elsewhere.

1   Many flightless beetles in Australia's northern wet tropics (a World Heritage Area) are restricted to a single forest subregion (Yeates *et al.* 2002). The large number of such species with presumed low dispersal ability in several families (namely Carabidae 86 species, Scarabaeidae 32 species, Tenebrionidae 87 species) implies that many, together with a variety of other beetles and insects of other orders (particularly Hemiptera: Aradidae), may indeed be vulnerable there as highly localized taxa.
2   Many of the dytiscid water beetles from underground calcrete aquifers in Western Australia (see p. 115) are known only from single aquifers which, following results of mitochondrial DNA investigations on the beetles (Cooper *et al.* 2002), may represent a series of subterranean islands with independently evolved beetle taxa.

The bewilderingly high richness of tropical beetle faunas, although long suspected, was brought to wide attention through Erwin's (1982) classic study of sampling beetles from the tropical forest canopy in Panama. His analysis founded later debate on the magnitude of tropical insect species richness. For the first time, Erwin provided testable hypotheses by which richness could be estimated and, although his assumptions have been challenged in detail, they have formed the foundation for considerable later evaluation (see Stork 1997). In acknowledging the massive diversity of beetles in tropical forests (as 'biodiversity at its utmost'), Erwin (1997) noted that they had been little used in interpreting environmental disturbance, for environmental monitoring or for understanding how tropical communities are structured, and also emphasized their great potential for augmenting our understanding of evolutionary biology and conservation. The sheer amount of information potentially available from tropical beetle faunas would have unique and massive importance in these areas of endeavour. The problems remain over how to harness and employ that information from such hyperdiverse groups and to overcome the current impediments to doing so. Much information on diversity emanates from studies on single sites or small regions, and the reasons for varying distributions and high beta diversity may be difficult to assess. Again from the Neotropics, only about 2.6% of the beetles of seven selected families from fogging samples were common to surveys from near Manaus (Brazil) and Tambopata (Peru) (Erwin 1988). These sites are separated by about 1500 km but, in a wider discussion of species turnover with distance across sites, Bartlett *et al.* (1999) noted that interpretations of distributions based on such separated samples are 'fraught with intensive site-specific differences that confound distance effects'.

The central paradox and values of beetles in conservation flow from their vast abundance and taxonomic and biological diversity. On the one hand, they offer abundant opportunities for study and evaluation of environmental changes. Almost every terrestrial or freshwater biome supports a wide taxonomic array of beetles, many of them responsive to some or other environmental change, whether natural or imposed, and many of them commanding attention as declining, either alone or as an entire specialized assemblage. The long-term interest in beetles noted earlier has laid a solid foundation of taxonomic and biological knowledge that aids some such appraisals, as well as suitable (ecologically informed) study and sampling methods. On the other hand, the bewildering variety of beetles is sometimes a barrier to understanding: we may indeed find numerous species in a locality, occupying collectively all or most trophic roles in a biome, but the detailed ecology of most (even all) the species is likely to be fragmentary, and the mechanisms sustaining them may need to be projected from little background other than from related taxa, or from similar biomes undergoing apparently similar processes or change. For conservationists, beetles, whether directly as conservation targets or tools for wider applications, offer both severe impediments and massive opportunity for progress.

In common with other insects with complete metamorphosis, conservation of beetles must consider the biology and needs of two very different life forms, whose ways of life may demand very different resources and conditions. Larvae and adults of the same species may coexist, or be separated in space and time, utilize different foodstuffs and occupy different feeding guilds. As Dennis *et al.* (2006, 2007) have emphasized for butterflies, successful conservation must determine and ensure the needs for both these active stages in a wider milieu in which the entire life cycle can be supported.

In much practical insect conservation and faunal documentation, high species diversity is a very mixed blessing. Relatively low-diversity groups, perhaps with only a few thousand species (however formidable such numbers seem to people used to working with mammals or birds), are fundamentally more tractable to non-specialists in particular. Within the insects, butterflies comprise only around 20,000 species, with regional faunas typically much smaller, and many of the genera and species are recognizable through well-illustrated field guides. A framework for their biology is also likely to exist, perhaps by reference to close relatives in the area or elsewhere. In contrast, with tropical beetles we are dealing with much larger numbers of taxa, most of whose biology and distribution is almost entirely unknown, and many of which are undescribed and undiagnosed. Many beetles may not be identifiable easily much beyond family level, and some to that level only with considerable difficulty attendant on small size and complex or confusing morphological characters. As Erwin, and many others, have emphasized, the decline of the taxonomic workforce has ensured that most of these taxa will not receive such formal treatment in the foreseeable future. In the terminology of Yeates *et al.* (2003), many beetle groups are 'taxonomically orphaned' by the absence of any specialist able to evaluate them in a regional or the global fauna, and comment on their affinities and peculiarities. Beetles are by no means alone in this regard: the situation may be even worse for parasitoid Hymenoptera for example (even in the best-documented temperate-region faunas; Shaw & Hochberg 2001), and paralleled in many families of Diptera and other diverse insect groups. Perusal of taxonomic journals (such as *Zootaxa*) in which numerous descriptive papers on beetles are published may convey misleading impressions: whereas many taxa are indeed being described, and many groups progressively revised, the size of the demand and the task ahead remain daunting.

Beetle recognition and identification

The practical ramifications of this absence of data are important. Lack of formal species' names and lack of ability to obtain those names (the situation sometimes termed the 'taxonomic impediment'; see Taylor 1983 for background) is exacerbated by high diversity to the extent that need for formal taxonomy cannot be fulfilled as a prerequisite for basic documentation, and is viewed widely by non-scientists as equivalent to lack of importance or interest. However willing and interested they may be, lack of taxonomic resources ensures that the few specialists on any individual insect group are substantially over-extended. Employer demands may effectively prevent such people participating in identification of ecological survey material for other people, and related activities. And, for groups such as beetles, the amount of material collected during such exercises can be formidable in both abundance and variety. It is one thing to ask a specialist to identify a single beetle or a few voucher specimens of particular interest or relevance to a study and within that person's sphere of interest (commonly a single beetle family or part thereof), but quite another to confront him or her with the entire outcome of a substantial survey, comprising perhaps hundreds of species across a wide array of families. Examining such collections is often a major research exercise in itself. However, lack of up-to-date guidebooks or other non-specialist publications renders such exercises almost impossible (and, at least, often unwise) to undertake without specialist direction and access to a major and well-curated institutional collection for comparison. Handbooks for many families are indeed available for many parts of the northern temperate zones in particular, and identification can then be undertaken with relative confidence, but it remains wise to attempt to have a series of voucher specimens checked by an experienced coleopterist, as recommended later for any survey in which the results may be used in recommendations for conservation management. The situation remarked by Crowson (1981), that few countries outside Europe have even reasonably comprehensive handbooks for beetle identification, although guides for particular families may exist, still pertains. Broad-based introductory illustrated handbooks to tropical beetles, such as those by Tung (1983) for Malaysia and Gressitt and Hornabrook (1977) for Papua New Guinea, are immensely valuable introductions to those faunas but can do little more than titillate for the wealth not included. Nevertheless, as Tung hoped, they can stimulate people to take up the study of beetles in such regions and lead to advances in knowledge. Clearly, the resources for the interested ecologist or conservation biologist to identify beetles easily and unambiguously beyond family level in much of the world simply do not exist, particularly locally. The practical dilemma is that for beetle faunas with high proportions of poorly documented species, non-specialist identifications are likely to often be erroneous, and specialists unlikely to be routinely available to help interpretation. Larochelle and Lariviere (2007), writing on New Zealand Carabidae, go further and state (p. 160): 'Species-based information should never be published or databased unless a carabid specialist has confirmed the validity of genera and species involved'. They noted also that isolated descriptions of new taxa, sometimes motivated by need to provide names for conservation targets, are misguided, and that beetle taxonomy should be pursued in the context of revisionary studies rather than piecemeal.

When, and if, a particular species of beetle is described formally is often serendipitous, and depends largely on the interest of a specialist examining that family or genus group at that time. Other factors also intervene: for the well-known British fauna, most larger beetles were described earlier than many small ones (Gaston 1991). The latter are commonly (i) more difficult to differentiate without close microscopical examination, or dissection of genitalic structures, and (ii) less attractive to many collectors and so less important unless with direct economic or other intrusive values. Gaston suggested that the smaller beetles may be simply less conspicuous, and harder to collect. This trend is by no means universal, as Allsopp (1997) found for the Australian scarabs, for which wide-ranging species were generally described earlier than many highly localized taxa, irrespective of their size. Many of the earlier-described species were those found closest to major human settlements (so that the south-eastern fauna was for long better documented than the fauna of the remote northern regions). However, importantly, all scarabs are at least moderately large beetles, and thereby reasonably conspicuous. Allsopp (1997) pointed out another possible anomaly relevant to conservation assessments – the probability that some of Australia's recently described scarabs currently have small defined ranges *because* they have been described recently, so that there has been little time to accumulate comprehensive information on their real distributions, which might be substantially underestimated from the material available. For Iberian Scarabaeidae, Lobo *et al.* (2007) also noted that mapping schemes (see p. 41) may show considerable bias, because initial records of species may be based on localities favoured by collectors seeking particular rare species and on more thorough exploration of places near to investigators' homes. Many hobbyists, seeking particular species but with limited recreational time available, will opt to visit traditional localities to seek their specimens rather than explore new areas that might not yield their targets.

Keys to beetle families are included in many general entomology texts, but regional bias may limit their usefulness. Most textbooks, for example, cater predominantly for one or other temperate-region fauna as their primary market and, at the least, the examples of beetles used to illustrate key characters may not occur widely elsewhere, or other faunas include additional families not treated in that text because of regional scarcity or absence.

Most information in texts, and most of the work referred to in this book, deals only or almost solely with adult beetles as the life stage most amenable to consistent recognition, easy collection and quantitative or semi-quantitative sampling. Yet, the major impacts of many beetles on human interests occur during the larval stage: the damage caused to pastures by subterranean scarab larvae, the variety of timber-boring larvae of several families, of leaf beetle larvae on crops and ornamental plants, the depredations of stored products pests, and so on. With the notable exceptions of some such pest groups or complexes, recognition of beetles to species or near-species level is best achieved on the adult stage, because many larvae have not been described, placed into a robust local taxonomic context nor associated unambiguously by rearing to the corresponding adults. Adult beetles in all parts of the world are documented more effectively than their larvae. As with any such bland statement on beetles, exceptions occur: excellent global keys to families and other higher groupings of larval beetles exist, with those to lower levels most developed for the northern temperate regions. Treatments such as that by Luff (1993) for larval Carabidae for a region of northern Europe are of immense value in recognition to finer levels of some relatively well-known groups. However, many, perhaps most, beetle larvae are not identifiable readily to species level, except by inference or clear association with likely corresponding adults. This situation reflects a point to be emphasized repeatedly, that the biology and life histories of most beetle species is incompletely known, so that constructive augmentation of this basic information is a common need in assessing conservation. Drawing on the wider literature on recognition of beetle larvae (from Boving & Craighead's 1931 survey onward), it may still be possible to detect particularly unexpected or unusual novelties in samples of beetle larvae but, in general, larvae have played little part in the development of beetle conservation studies. Whilst recognition of larvae underscores the integrity of much applied ecology of economically important beetles, equivalent importance has yet to be facilitated for conservation.

As implied from Larochelle and Lariviere's (2007) comment above, decription of a species, either adult or larva, does not alone convey unambiguous recognition. Isolated descriptions of beetle species (many of which, particularly until the early decades of the 20th century, were brief, unillustrated and based on character suites that are by more modern standards regarded as superficial or inadequate) may not provide adequate comparison with close relatives or, if based on few specimens, may not recognize individual variability. Published dichotomous keys encapsulate diagnoses of known species, but users must be aware of their limitations and some possible caveats on their uncritical use. As one common example, discovery of additional species renders any such key incomplete, so that species may be forced spuriously into the best available category (name) by a non-specialist user, even if it is abundantly distinct. Some groups of beetles are much more prone to this augmentation than others, and it is useful if compilers indicate possible problems by, for example, (i) noting the relative likely completeness of the material used (are there likely to be many unincluded species or is the key based on reasonably comprehensive appraisal?) and (ii) noting if particular key couplets are unsatisfactory or indicate possible or actual complexes of species rather than solely the name provided. Zimmerman (1994) prefaced some of his weevil keys with a comment to the effect that the keys were simply to separate specimens examined, not necessarily all the species. Traditionally, taxonomists, of beetles or other organisms, have tended to write for their peers; in conservation the needs for unambiguous species recognition and detection go far beyond that clientele, and need to be understood by people with little specialist knowledge of the insects.

Not unexpectedly, some beetle families are better known taxonomically and biologically than others. In particular those families containing larger and more conspicuous or colourful adult beetles have long been popular objects for collectors, and the philatelic aspect of beetle collecting has stimulated their study and provision of names. In recent years, it has also led to production of many well-illustrated (by paintings or photographs of all available taxa) books (many of them expensive, and of varying quality and scientific value) on selected families, such as Carabidae, Lucanidae, Cerambycidae, and larger scarabs, to cater for this interest. As indicated above, size matters to many collectors, and the largest and most spectacular beetles are commonly also those easiest to identify and also those for which accessible identification guides are most likely to be available. The black holes in beetle taxonomy and the greatest challenges for faunal interpretation are indeed mainly within the smaller beetles: the insect equivalent of the little brown bird of ornithologist 'twitchers' is assuredly the little black beetle! However, even a beetle in the hand may be extremely difficult to allocate easily even to family level. Most small members of many beetle families are not sought by most general collectors and remain the province of a few specialists, but also may have far greater economic importance than some of the more charismatic taxa, so that recognition of species has major importance in designing pest management: again, scientific study has a basis in need to understand those taxa, either as pests or as beneficial species. The latter emphasis collectively includes considerable variety, reflecting that particular beetle families or subfamilies may have very distinctive ecology as herbivores, predators, fungivores, detritivores (or other trophic category) and collectively exploit any available foodstuff in terrestrial or freshwater ecosystems.

However, whereas beetles participate in virtually all ecosystem processes, limitations in species-level taxonomy pose two main restrictions to evaluating these in many places. First, that taxonomy of most beetles is adequate for confident appraisal by non-specialists only in a few places, such as the UK. Second, some families are far better known than others, and wider geographical appraisal is thus feasible only for such biased subsets from the order. Notwithstanding this, individual beetle species from many families have aroused concern because of perceived declines and apparent threats to their well-being. Usually, these must be considered as isolated cases rather than as members of a well-studied fauna. It may indeed be a formal requirement for such species to have scientific names as a condition of listing for conservation interest, so compounding the dilemma noted earlier.

In addition to their ordinal diversity and the presence of individual notable species, beetles have three major advantages over most other animals in ecological projects and those directed at environmental assessment of various kinds.

1 Many of the more diverse families can be characterized easily to that level. Even if the individual species cannot be named easily, the general appearance

allows for easy and largely unambiguous recognition of a ground beetle, rove beetle, scarab or weevil in samples. The insects present can subsequently be allocated to consistently recognizable categories such as morphospecies to enable some level of quantitative analysis that does not rely for its integrity on full species-level taxonomy. Many such families are distributed widely, have a broadly definable trophic role in communities, and contain numerous species. They are thus informative to ecologists, as beetles may be available for studies of almost any ecological role, process or interaction, either directly or as surrogates.

2   Methods for collecting or more formally sampling beetles are well established, and many of them rely on cheap, easily obtained and easily transported equipment. Most methods developed by hobbyists can, at least to some extent, be modified easily to form the basis of rigorous sampling with varying levels of standardization and replication needed to afford quantitative or semi-quantitative interpretation. The variety of traditional collecting techniques is limited only by the ingenuity of the practitioners. Background literature on the limitations and caveats attendant on most methods is extensive, and each method has its devotees, its detractors and its values in particular contexts, which need to be defined and understood. Many beetles can be reared easily, so that larvae and adults can sometimes be associated clearly from multispecies samples. However, many large beetles have long lifespans (see p. 146), so that the time available for a survey may not allow this exercise to proceed.

3   Specimens can be processed or prepared for study or exhibition easily, a factor that has contributed to their popularity as collectable objects. The importance of synoptic collections or well-prepared and well-documented voucher specimens of all species (or morphospecies) found during a survey or other study cannot be overestimated. They are the essential points for future reference and comparison, and for the validation of identifications as taxonomy progresses in the future.

Sampling and surveying beetles for conservation

In conservation or other ecological survey assessments, sampling beetles may be undertaken for a variety of purposes. Broadly, beetles may be used to illustrate ecological patterns in space and time, and also be important functionally in ecological processes within any biotope or more broadly. As individual targets for conservation, information on individual beetle species may be a priority. As tools in wider assessment, primary emphasis may shift to wider aspects of beetle assemblage composition and its changes. These approaches thereby include the following.

1   Inventory: an attempt to produce a list of all taxa occurring in an area or habitat, or in association with a particular resource such as a specified plant species. It may be necessary to employ a considerable variety of approaches to increase comprehensiveness of sampling for a more complete inventory,

or to focus very specifically to exclude unassociated species from a resource. At either scale, the intention is to accumulate the greatest representation of the species present as a measure of richness that can reflect the importance of the sampling arena and, perhaps, serve to compare or rank it within wider comparisons either across sites or across habitats within a site. For any comparisons, sampling methods and effort must be standardized across sites or occasions. The most informative inventories include samples taken over a sufficient period (of at least several intervals throughout a year) to accumulate seasonally apparent taxa. Many beetles are either present for parts of each year or have seasonal periods of activity and consequent amenability to trapping. For example, some carabids in Tasmanian eucalypt forests may be present for most of the year, but their activity may vary substantially at different seasons, so that their representation in trap catches varies considerably with season (Michaels & McQuillan 1995). The outcomes of an inventory survey are presented most usually as species lists, with or without measures of relative abundance of the taxa represented. Completing such inventories is a lengthy process, of course: for beetles in his home garden in central England, Welch (1990) kept records to accumulate around 760 species over 16 years, with new species found every year. For perspective, this total represents almost 20% of the British beetle fauna. In contrast, most practical inventory studies can be undertaken for only limited periods. A similar approach may be used to evaluate the richness of particular taxonomic groups, such as major families, or feeding guilds in an area.

2 Detecting the presence of particular signal species, such as those known to be threatened or otherwise of conservation interest, or the establishment of introduced species such as biological control agents. For such exercises, it may not be appropriate to kill specimens, so that the spectrum of techniques is restricted to those that are non-destructive. Particularly for threatened species, quantitative data (although valuable) may be secondary to simply determining presence at a site as a basis for pursuing conservation. The converse, proving absence of such a species (especially on sites where it has been known previously), is much more difficult. A reasonable basis for this is that sufficient effort has been made to detect it that it is likely to have been discovered if it indeed occurred there. Traditionally, many beetle collectors have heeded the maxim of deliberately seeking particular rare or elusive species rather than general collecting, as the more common species are then likely to be collected anyway (see Walsh & Dibb 1954 for background). With such an aim, seeking all available biological and distributional information beforehand to aid effective searching parallels the needs for any conservation survey.

3 Single-species studies may expand to fuller autecological studies, undertaken for a variety of purposes, commonly including aspects of seasonal development, population dynamics and distribution. These are based on sequences of sampling events over time and perhaps replicated across sites. In addition to studies undertaken to help understand the biology of species of conservation interest, common contexts include (i) understanding the dynamics of pest species as a foundation for designing management and (ii) monitoring

the presence, establishment and impacts of predatory species used in such management. Extensive surveys may also be needed to monitor effects of management of threatened species and to modify management in response to the findings.

4   Allied to the above, studies on species of conservation interest sometimes address impact of particular threats or suspected threats, either by correlation with different ecological regimes in which the purported threat is displayed, or by direct causal investigations. The latter can be more complex for rare species, simply because sufficient numbers may not be available for replicable manipulative experiments, particularly without risk of causing further harm.

5   Studies of assemblages may incorporate the need to monitor any imposed change, so that the focal beetles are used as an index of diversity or other form of indicator to provide wider ecological insight or inference. Changing species richness or composition can reflect environmental and seasonal conditions, so that inter-site comparisons must be based on samples taken at the same time of year, by the same method and comparable sampling effort, and avoiding any other sources of unplanned variation that render such comparisons dubious. Seasonal patterns of beetle apparency, whether based on life cycles or suitability of the local environment, are very varied. In temperate regions, some can be highly predictable. Particularly in the tropics, many insect species exhibit complex patterns of seasonality (Wolda 1988) and even in relatively aseasonal environments there, some beetles can be highly seasonal. As demonstrated later, evaluation of beetle assemblages may need to consider altitude, climate, vegetation type and seasonality each as a key influence on composition, in addition to issues arising from sampling techniques and analysis.

Whatever the primary purpose of a conservation-oriented field study on beetles, the scale of the study must be considered realistically. The extent and causes of changes in beetle abundance and distribution may reflect local environmental factors (such as microclimate or individual host plant condition) and/or larger-scale influences such as fragmentation in the landscape (see p. 92). Heterogeneity in distribution of species and assemblages can be unexpectedly high, with different patches of agricultural land each supporting local assemblages that differ in detail from those on apparently similar patches in their vicinity, but which draw variously on the same regional pool of taxa. Example studies in which various ecological scales have been incorporated include carabid beetles on farmlands (Kinnunen *et al.* 2001) and bark beetles in forest stands (Peltonen *et al.* 1998), both in Finland; there are many others.

At this stage, it is also appropriate to note an important aspect of interpreting assemblages arising from lack of taxonomic information (see p. 68), namely the trend to use higher-level taxonomic levels, rather than species, for analysis. We read repeatedly of family X or genus Y as entities in insect surveys, either as components of richness or as indicators (see p. 47). Use of such larger entities saves massively on costs of analysis, and is logistically very attractive because of this. However, it is commonly fundamentally inadequate to provide the information desired as an objective of the surveys undertaken, because it masks the detail afforded by species-level studies. A compelling case for species-level appraisal of assemblages and diversity (Spence *et al*. 2008, echoing similar sentiments from others over recent decades) has emphasized yet again that individual species differ in ecology and functional role, genetic constitution and conservation need. The authors' argument, which is paralleled in numerous surveys in which beetle families or genera are the taxonomic levels employed for interpretations, sometimes leading to far-reaching management decisions, illustrates some limitations. Spence *et al*. suggested the parallel that ornithologists would dismiss conservation-oriented appraisals that grouped the several ecologically different species of North American jays as corvids, and noted that even generic groupings of birds (such as *Corvus*) would probably be discounted or ridiculed (as would groupings such as parrots or honeyeaters in Australia, or finches in Europe), simply because these groups, however valid as descriptive entities, are sufficiently well known to reveal the major differences between constituent species – it also shows the absurdity of lumping such biologically disparate taxa together. As Spence *et al*. (2008) noted, insects are no different to vertebrates in this regard. Many families and genera of beetles manifest massive ecological variety at species level, and there may be little value in obscuring this by shallow or uncritical taxonomic penetration: inventories for conservation evaluation may be of little value unless constructed at the species level, with the practical proviso that consistently recognized morphospecies may need to be employed rather than taxonomic species in many places. Morphospecies maintain an equivalent level to species for richness estimations and consistent focus for compositional changes, but their persistent value depends on responsible deposition of voucher material (see p. 16). Without this, many entomologists are perhaps guilty as charged in suggesting that species-level work for arthropods is scientifically essential when it is easy, but discretionary when it is difficult to achieve! Wherever feasible, taxonomic analysis should be viewed as an important facet of increasing the value of assemblage descriptions, changes or functions.

One of the more comprehensive surveys of tropical beetles was of weevils (Curculionoidea) in Panama, where Wolda *et al*. (1998) used light traps at seven sites (from sea level to 2200 m), with climates ranging from sharply seasonal to almost aseasonal, and spanning a variety of habitats (from natural tropical forest to highly altered areas). Altogether, 2086 species were accumulated in sampling over three consecutive years at each site (collectively spanning 1976–85), and this was considered to be only a small proportion of the species present, because many weevils are not attracted to light. Species ranged from those with very strict seasonal incidence to those occurring year-round, with those at climatically seasonal sites tending to be most abundant at the start of the rainy season. As noted earlier, pronounced seasonality is common in beetles, and particular species may exhibit very characteristic patterns of appearance. In temperate regions, for example, some species of tiger beetle emerge only in midsummer, others only in autumn and hibernate before reproducing; in drier regions adult emergence may be associated with onset of seasonal rains. Trying to predict any seasonal pattern in a poorly known fauna may be difficult. Use of information compiled from specimens in museum collections may sometimes be useful, but may reflect the seasonal activity of collectors as much as that of the beetles! One example is for riverine tiger beetles in South Africa, for which most collection records are from the rainy season but, from additional systematic survey at other times of the year, the beetles are suspected to be active for much of the year (Mawdsley & Sithole 2007).

For the above purposes, and the numerous intergrading situations that can arise, several survey techniques are commonly used. Many of these are so-called passive techniques that trap and in many instances kill beetles for later examination. These cannot be deployed in studies where any such mortality may be harmful or undesirable, as for studies on rare or threatened species, and must be curtailed immediately if known threatened species are unexpectedly found in catches on sites where they were not previously known to occur. Preliminary investigation may be wise, and certainly responsible, in cases of doubt over such occurrences. In some conservation exercises, it may be necessary to modify standard methods to capture the beetles alive, for example for translocation exercises. As one such modification, Weber and Heimbach (2001) used pitfall traps with floating cork 'islands' to prevent carabids drowning in wet weather.

A selection of beetle sampling techniques is summarized in Table 1.1. Whatever approach is used, care must be taken not to damage the habitat, not to over-collect or over-sample, not to have untoward side effects such as excessive bycatch, and not to transgress the conditions of any permit or access conditions, and to ensure that the material taken is treated responsibly. All too often, massive numbers of invertebrates are captured indiscriminately and killed during surveys, and many are never examined or analysed critically. Many are regarded as bycatch (representing non-target groups) and may simply be discarded. At the outset of any survey for beetles, pragmatic decisions must be made over the resources needed for sample interpretation and processing of specimens, in particular which taxa are to be used, and at what taxonomic level. It is very easy to collect far more material in an insect survey than can be appraised realistically during the planned life or budget of that project. The limitations of any method used in relation to the purpose of the study must also be understood clearly; for example, some species of Carabidae are underestimated in pitfall trap surveys (Halsall & Wratten 1988), so that using this technique alone may misrepresent information on the supposedly sampled assemblage. Full details of the methods used should be included in any report or publication flowing from the study. Many variables affect trap catches and efficiency, and simply listing and assessing these is an important component of assessing trapping effort. Thus, although baited pitfall traps are a standard tool for trapping dung beetles, only recently have serious attempts been made to determine the capture arena of such traps by investigating distances over which beetles may respond. Mark–release–recapture trials with the small scarab beetle *Canthon acutus* in Venezuela suggested that traps should be separated by at least 50 m in order to be considered independent (Larsen & Forsyth 2005), a distance far greater than the 5–10 m spacing commonly employed with presumption of trap independence in surveys.

Some techniques have become standards for particular beetle groups, and are used almost universally in their study. For example, pitfall trap surveys for ground beetles and dung beetles have yielded a high proportion of the specimens used

Table 1.1 Summary of some sampling/collecting methods for beetles and used in surveys for conservation studies (see text for examples including many of these).

Method	Principle and targets	Variables and conditions
Pitfall traps	Containers sunk in ground Surface-active beetles fall in Non-selective	Size (diameter), spacing, duration of exposure, use of drift fence Preservative and baits may be included Easy replication Can be roofed Semi-quantitative
Tullgren funnels	Samples of leaf litter heated and dried from above Beetles fall into jars of alcohol or other preservative Many small taxa collected Use also for soil cores	Dry loose litter best Volume, duration Need power supply Semi-quantitative One to a few days extraction time
Winkler bag	Samples of litter sifted and bagged Hung so animals drop, as above Based on animals moving to reach shelter rather than responding to heat	As funnels, but no power needed
Litter sifting	Direct inspection of litter in field, by sieving or sorting Yields many small beetles from riverine litter/debris or other restricted habitats	Based on volume or ground area samples for replication Can be time-consuming and laborious

Table 1.1 Continued.

Method	Principle and targets	Variables and conditions
Direct netting	Use of typical butterfly net to capture individual beetles in flight or on the ground	High selectivity in capturing voucher specimens
Sweep-netting	Strong net used to 'swish' repeatedly through low vegetation Dislodged beetles removed individually from net	Vegetation must be dry Time of day and weather may be influential Semi-quantitative rapid method, can be area based for replication
Beating	Canvas tray held horizontally under low tree/shrub branches Branches struck sharply with stick, dislodged beetles collected on tray (many may not move, and be cryptic)	Vegetation must be dry, as above Can standardize search times for collecting from tray
Suction sampler	Vacuum cleaner used to collect insects from low vegetation into container	As above Can be applied to individual vegetation units May accumulate much debris and require further sorting
Insecticide 'fogging'	Mist blower used to 'fog' tree canopy with pyrethrin insecticide, catching falling insects near ground, in funnels or on trays Many beetles found only in canopy layer	Forest canopy otherwise inaccessible to sampling Can be very specific by sampling individual trees Considerable time needed to set up Equipment heavy
Bait traps	Variety of procedures involving use of attractants, mainly for flying beetles, in conjunction with pitfalls or other retention device	Specific cases include use of dung (Scarabaeidae), carrion (Silphidae), fruit or pheromones Considerable variety of uses
Trap logs/wood	Placement of cut or fallen logs or billets, and later examination to rear or collect timber-infesting and bark beetles	Comparative values of different woods: host specificity trials May take a year or more to gain information needed

Method	Description	Notes
Light traps	Attraction of flying insects to ultraviolet light Collect in container or when resting on white sheet or background	Weather, time of night, phase of moon, season, etc. all important influences
Malaise trap	One of several patterns of intercept trap Flying insects contact vertical fine black mesh barrier, being directed upwards into container	Passive, acting by day and night High bycatch
Window trap	Intercept trap for flying insects Contains vertical panel of glass, mesh or perspex, from which insects drop into trough of preservative	As above Best for larger beetles Can be roofed to prevent flooding
Emergence traps	Enclosure, often of mesh, vegetation, wood or other substrate, with provision to capture insects emerging from it, such as by funnelling upwards into container	Can be used in aquatic or terrestrial habitats
Eclector traps	A form of emergence trap, usually a plastic or other funnel attached to an area of bark or wood to catch emerging beetles and others	
Dip-netting	Use of, usually, triangular long-handled net in water to capture insects from water column or amongst submerged vegetation Variety of small, otherwise elusive, beetles	Can be used in standard way, as aquatic sweep net
Direct searching	Often very rewarding in yielding species not otherwise obtained easily Can combine with sifting	Universally applicable Often a valuable adjunct to other sampling methods, and can help to provide more specific information on habits and associations Use by day or night (with head-torch)

to interpret species assemblage compositions, changes and distributions. However, use of any particular trapping method must be tailored to local conditions and faunas. Within the Hawaiian Carabidae, for example, Psydrini are best sought by a suite of direct searching methods, and Liebherr and Zimmerman (2000) made no mention of pitfall trapping in their commentary on this substantial archipelago fauna. Beetle trapping methods can sometimes be standardized within narrow limits for evaluating sampling effort. Any single method used in the belief that it is indeed the best can still not usually secure all species present, so that for inventory studies some form of sampling set (or combination of different methods selected to be complementary) is wise. Thus, Davis *et al.* (2001) used baited pitfall traps and flight intercept traps to assess richness of dung beetles in Malaysia. Both methods draw on normal beetle behaviour in dispersing to find and exploit dung. Collectively, the two methods yielded 35,279 beetles, representing 86 species. Application of species richness indices to predict numbers of species (Chao 1 and Chao 2) gave 78 and 79 species from the pitfall data and 88 and 85 species from the flight traps, implying that the latter method may provide a more comprehensive evaluation of beetle richness. In this example, the accumulation could be evaluated against a much fuller one available for the area (97 species from 68,481 individuals sampled) to assess its representativeness. Often no such background resource exists to provide realistic perspective, and richness indices or sampling accumulation curves may then be useful estimators of sampling adequacy. In another informative survey, Larochelle and Lariviere (2007) individually noted the techniques preferred to collect each genus of Carabidae in New Zealand. Altogether, about 25 collecting approaches were noted. The most frequently cited were (i) pitfall traps (for 54 of 86 genera in the fauna) and (ii) turning over of fallen trees, logs, stones and other ground materials (for 52 genera, and reflecting high incidence of such cryptozoic taxa). Several ecologically specialized taxa demanded correspondingly more specialized approaches to retrieve them, but such detail may become apparent only after considerable survey effort and experience. Studies demonstrating the different spectrum of beetle species from different trapping methods abound. As another example, comparison of boreal forest beetles captured in Finland by window traps and pitfall traps showed 62 of the 435 species only in pitfalls, 250 only in window traps and only 123 species in both trapping regimes (Simila *et al.* 2002). Again, the relative merits of several approaches to sampling saproxylic beetles were compared by Alinvi *et al.* (2007), who suggested a combination of window traps and eclector traps as highly complementary methods. Window traps sampled the local species pool by intercepting the beetles attracted to dead wood, whereas eclector traps provided more detailed information by capturing the beetles emerging from particular pieces of wood. Of the 148 beetle species captured in Sweden by these methods and bark sieving, only 22% were taken by more than one method. The differing arrays within the four predominant families (Table 1.2) indicate the magnitude of differences that can arise in assemblage data based on only one sampling method, and that cannot itself be evaluated without comparative studies.

   More generally, beetle surveys undertaken in underexplored areas or faunas may depend initially on methods well tried elsewhere, and development of

Table 1.2 Relative representation of four dominant families of forest beetles taken by three sampling methods from logs of spruce (*Picea alba*) in Sweden. Values are given as percentage of species or individuals captured, with actual numbers in parentheses.

	Method		
	Eclector trap	*Window trap*	*Bark sampling*
Number of species			
Carabidae		6% (7)	
Curculionidae	31% (9)	11% (13)	23% (10)
Leiodidae		11% (13)	
Staphylinidae	28% (8)	52% (60)	43% (19)
Number of individuals			
Carabidae		8% (65)	
Curculionidae	69% (100)	6% (51)	74% (474)
Leiodidae		12% (97)	
Staphylinidae	18% (26)	67% (536)	18% (112)

Source: Alinvi *et al.* (2007) with permission.

comprehensive or locally informed sampling sets not be easy. Different sampling methods may differ less obviously than in their broad approach, with small differences in design affecting the catches markedly. Different baits in pitfall traps, for example, may yield different spectra of dung beetle species, so that Larsen and Forsyth (2005) suggested baiting traps with vertebrate carrion, invertebrate carrion, rotting fruit and rotting fungus as well as with 'basic dung'. Lack of standardization of method details renders comparison of the results from different beetle surveys very difficult, even though very similar basic methods are employed.

However, specialized investigations sometimes require considerable ingenuity and inventiveness to devise suitable techniques, so that novel sampling methods for beetles abound, and continue to be developed. For example, a backpack vacuum cleaner was used to extract debris (including beetle larvae and remains) from deep hollows in old standing trees (Bussler & Müler 2009), with living larvae replaced after identification. A key focal species in that study, *Osmoderma eremita* (see p. 165), is viewed as a surrogate for wider richness of beetles in this habitat (Ranius 2002). In a further innovative sampling approach, Svensson *et al.* (2003) demonstrated the values of sampling air in the tree hollows to detect male *Osmoderma* beetles by presence of their sex pheromone (R-(+)-γ-decalactone) by gas chromatography and mass spectrography. The chemical, giving the beetles a characteristic peach-like odour, appears to be a reliable indicator of beetle presence, but a limitation is that its absence does not necessarily mean that beetles are also absent, as it dissipates quite rapidly. A somewhat similar approach using attractants for beetles was discussed by Chapman *et al.* (2002) for *Lucanus cervus* (see p. 198), as a basis for long-term and non-destructive population monitoring. Pheromone sampling is widespread for some pest beetle

monitoring. Fireflies (Lampyridae) can sometimes be estimated by simply count-
ing flashes over a given interval, with Yuma (2007) reporting counts of Genji
firefly in Japan (see p. 2) over 25 years, and calibrating counts by comparison
with mark–release–recapture assessments. Cerambycid infestations in wood may
even be detectable by the sounds made by larvae, with this approach examined
recently for the economic pest Asian longhorn (*Anoplophora glabripennis*,
Cerambycidae) in North America (Mankin *et al.* 2008). Use of acoustic techno-
logy has been pursued for this species as a possibly more satisfactory alternative
to current laborious physical inspections of trees, but determining the signal profiles
and distinguishing them from background noise remains difficult. Mankin *et al.*
noted, for example, the difficulties of detecting beetles during high winds or in
high traffic noise.

Many sampling methods for beetles have been investigated most critically for
common beetle species, such as agricultural or forest pest species, for which detailed
information on dispersal and behaviour may be important in management. These
contexts may aid more critical focus for species of conservation interest and indeed
many others in similar biomes or landscapes. Even for common beetles, a novel
or previously untested approach may lead to major revision of the conventional
wisdom of ecological knowledge. The bracket fungus-infesting *Bolitophagus reticulatus* (Tenebrionidae) in Europe was long believed to have very low dispersal
capability, reflected in low trap catches across several studies. This inference was
challenged by results from investigation of the attractant effects of volatile chemicals (Jonsell *et al.* 2003), demonstrating both high attraction of flying beetles
and an intense but short major flight season. The beetle appears to disperse
sufficiently well that early concepts of it forming metapopulations (with individual
Fomes fungi the component units) may not be correct, and the population in
a forest may indeed be continuous. For some forest beetles, distributions may
be evaluated through use of aerial photography to detect changes in tree condition, with this approach now becoming highly refined since earlier realization
that particular spectral bands may provide fine-scale estimations. For the seriously
damaging southern pine beetle (*Dendroctonus frontalis*), early attack could be
detected by changes in tree colour through chlorosis. Carter *et al.* (1998) used
pixel sizes of 1×1 m ground areas, so that individual trees could be assessed for
condition from photographs taken from a flight altitude of 1830 m. Although
not a conservation survey method, this, now early, example indicates some of
the potential of remote sensing techniques for beetle surveys, and several more
pertinent approaches are noted elsewhere.

The above categories of survey are all predicated on beetles being the primary
focus of the survey exercise. However, discovery of unusual beetles during surveys directed initially at other taxa, or in more general appraisals, may elevate
their priority from these and lead to more targeted study. Assemblage studies
have sometimes arisen from more general insect studies, such as use of pitfall
traps for ants, that commonly also trap numerous beetles as bycatch (see p. 121).
It may then become important to refine the initial sampling regime for greater
effect in assessment, or to protect such taxa.

The sampling regime must reflect the precise questions being asked, so that
defining the objectives of any beetle survey before it is started is important.

Table 1.3 Activity distribution amongst dung beetles surveyed by baited pitfall traps in French Guiana illustrating occurrence of guilds of species with different daily activity patterns. The pool comprised 63 species (with six species not categorized), and richness and relative abundance are shown, with percentages in parentheses.

Activity pattern	No. of species	Abundance
Diurnal	27 (42.9)	901 (33.8)
Nocturnal	13 (20.6)	431 (16.2)
Dawn and dusk active	14 (22.2)	367 (13.8)
Nocturnal–diurnal	3 (4.8)	934 (35.1)

Source: Feer & Pincebourde (2005) with permission.

Contexts such as those noted above must be appraised in relation to the scale of sampling needed and the resources available for an ideal programme to go ahead, sufficiently planned and replicated in time and space where relevant. Many demands for conservation surveys, for example, are for one-off investigations without provision for comparisons over time or across sites, and they may have severe limitations for providing definitive information. Possible questions of scale include whether a notable species is to be sought at a single site, across a series of sites within a given region, or in other likely areas across a wider inferred or historical range. Likewise, are quantitative data or ecological knowledge needed and, if so, what are the projected uses and analyses for that data? Another context is whether the study forms part of a larger endeavour and needs integration (or predetermined sampling approaches) based on that but initiated elsewhere. Such considerations are easy to overlook, but emphasize the need for very careful thought and experimental design (sometimes involving collaboration with other scientists or agencies) early in a survey plan, not least to facilitate costing of the exercise and assuring the needed logistic support. Perhaps the most important decision is how to ensure that sufficient sampling is undertaken, by the most suitable methods, to answer the questions posed adequately, whilst not incurring extra costs by undertaking work that is not needed. For more general objectives, a suite of more general methods may be needed; for more targeted questions or single-species studies, these methods may need more careful fine-tuning in relation to the species phenology and biology. Short-term or one-off sampling may even need to consider patterns of diurnal activity of beetles, with many species active (and thus trappable) only at particular times. Differences constitute one important aspect of ecological segregation, common in many different insect groups. Daily flight activity patterns of Scarabaeinae in French Guiana revealed distinct cohorts of diurnal, nocturnal and crepuscular species (Feer & Pincebourde 2005) (Table 1.3), with the first about twice as rich as either of the other guilds. Distinct diurnal and nocturnal separation of dung beetle species has been documented quite extensively in several parts of the world. Feer and Pincebourde found two distinct patterns among diurnal flyers: some species flew predominantly in the first half of the day, while others flew throughout the day. Most nocturnal species flew during the first half of the night; some crepuscular species flew at both dawn and dusk, but others only at dusk.

Greater dominance by diurnal species may reflect greater dung deposition by mammals during that time, with the reverse more likely in Australia where most tropical mammals are nocturnal.

Studying rare species

Many rare species, including almost all the beetles scheduled for conservation attention, are particularly difficult to study and survey, simply because they occur in very small numbers and very low densities. In addition they should not be sampled by any technique that might cause loss or harm to individuals or populations. Even experienced specialists searching intensively for taxa with which they are familiar may find few specimens of their target group. Bell (1985) noted that he spent 3 months at Wau (Papua New Guinea) during which he searched for a group of carabids (Rhysodini) in logs. He described these as 'uncommon, secretive beetles, spending most of their lives within decayed wood'. Bell found only eight logs with these beetles, collectively yielding 37 adult beetles representing six species. Experimental manipulations of such 'genuinely rare and elusive' species may be impossible, and need for statistical analyses may dictate particular sampling approaches to provide the data in suitable form. This problem has been addressed for red-listed saproxylic beetles in Sweden (Hedgren & Weslien 2008). Two sampling regimes involved random sampling (selecting position by GPS within stands of spruce and sampling the nearest dead tree) and subjective sampling (dead trees in the same stands selected on available biological knowledge as being those likely to host the beetles). The latter approach was significantly more efficient (red-listed beetles found in 28 of 78 subjective samples, compared with 56 of 360 random samples). However, both series yielded a substantial set of these rare species (12 in subjective samples, 13 in random samples, with a combined pool of 17 species). The method preferred may be dictated by the aim of the study: subjective sampling may be more cost-effective for rapid surveys to determine conservation value of a stand, whereas random sampling may yield new knowledge and provide data more accessible for formal analyses because it is more objective and easily replicated; it may thus be preferred for purposes such as long-term monitoring.

Because of the unpredictability of finding rare beetle species in samples, rendering the enumeration of these taxa very uncertain, Martikainen and Kaila (2004) suggested that reserve selection based on these species should be cautious. In their 10-year survey of beetles in birch-dominated forests in Finland, 258 of the 583 species captured were saproxylic, but many of the rare species were not found until after several years of investigation, and most were then seen only in very small numbers. For example, the only individual of the vulnerable *Phytobaenus amabilis* (Aderidae) was taken in year 6, and that of the endangered *Neomida haemorrhoidalis* (Tenebrionidae) in year 8. Such species indeed appear to be extremely scarce, and only four of the 16 species of individual conservation concern yielded more than 10 individuals over the extended sampling period. Accumulation curves for species of saproxylic beetles in the two forests sampled (Fig. 1.2) revealed that a high proportion of the common species

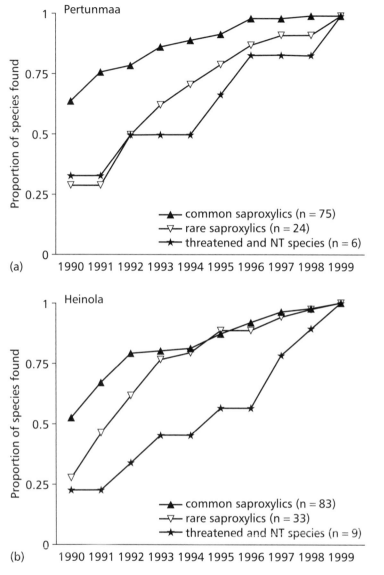

Fig. 1.2 Accumulation of species richness of three categories of saproxylic beetles (common, rare, and recognized threatened and near-threatened species) in two birch-dominated forest sites in Finland: (a) Pertunmaa; (b) Heinola. (From Martikainen & Kaila 2004 with permission.)

(those represented by 51 or more individuals in total) had been detected after 2–3 years of sampling. Accumulation of rare species (less abundant than the above) was much slower, with sampling asymptotes not reached after 10 years. The accumulation of threatened and near-threatened species appeared highly incomplete, with additional species still accruing at the end of the survey period. In short, even after this lengthy survey, the number of saproxylic beetle species in the forests remained unknown. Martikainen and Kaila recognized the possibility that the resident fauna had indeed been sampled adequately, and that the newer records were of vagrants from other habitats. They argued that, should this be the case, those species should perhaps be common in other habitats and so more common in the samples, and suggested that the ecologically specialized nature of many of the more recently accumulated rare saproxylic beetles might indicate resident species that are not detected easily. As with some other studies, rare species were sampled comprehensively only with considerable difficulty, and the uncertainties over what factors influence their incidence and abundance render reliable comparison of different areas or habitats very tentative.

It is obvious, though not always acknowledged openly, that any survey can benefit from knowledge of the taxa sought, so that redundant sampling effort can be avoided in both space and time. This knowledge becomes particularly important when targeting particular species, to help avoid wasted effort, but may also apply to assemblage studies. Writing on South African dung beetles, for example, Davis (2002) noted that failing to survey particular local habitats and dung types may result in absence of records for many species. Numerous dung beetle species around Pretoria were extreme specialists on sand, and some were recorded only from particular soil or vegetation types. Seasonal variations in appearance also occur. More generally, most dung beetles in the region are characteristic of particular ecoclimatic regions, some of them constituting areas of substantial endemism. Similar patterns are found in numerous other beetle groups.

Especially suitable habitats for beetles may be very small, and dispersed widely in a landscape. Actual critical sizes of habitat patches are difficult to assess but are of vital importance in considering values of fragments (see p. 94) or small 'island' habitats (see p. 109). Likewise, small habitats may be difficult to detect in complex landscapes. However, with a sound framework of what characterizes good habitat (see p. 77) for particular species, remote sensing approaches may have value. For tiger beetles, Mawdsley (2008) used two web-based systems (Google Earth, Microsoft Terranova) to help locate small patches of potential habitat in complex landscapes, and considered that the approach 'shows great promise', especially for landscapes where visual contrasts may indicate suitability. Many important variables, such as soil salinity or organic content, may be important for tiger beetles and cannot at present be appraised by this approach.

For inventory studies, a high proportion of the species retrieved are likely to occur in very small numbers, many of the beetles by singletons, and thus be rare in sampled assemblages.

The major dilemma for interpretation arises with the realization that the simple detection of such species targeted for conservation attention may require very considerable sampling effort, but those species may be the ones of major interest from the viewpoint of species conservation, with conservation interest enhanced by supposition of rarity. They are thus those for which quantitative information may be particularly valuable. Enormous numbers of beetle species must at this stage be considered rare, simply because they are known from only single specimens or very few individuals, and from single sites or samples. Many have been the subjects of targeted surveys that have proved futile, but it is often difficult to determine whether a species is genuinely rare or simply not retrieved. Klausnitzer (1983) noted a species of *Rhipidius* (Ripidophoridae), then known from only one European specimen caught in 1867 with a second specimen found in 1929, as 'probably the rarest beetle in Central Europe'. Rarity is by no means confined to small or obscure beetles; one of the world's largest species, the South American *Titanus giganteus* (Cerambycidae) was for long known only from very few specimens, and its detailed biology remains undocumented.

A typical sampling pattern will continue to yield additional low-abundance species with additional sampling, so that the number of rare species in an inventory reflects sampling effort, and will increase knowledge of the richness of any local fauna. Typically, increased sampling effort will also lead to increased representation of the few abundant species, and continually add to the 'tail' of scarce species. Boreal forest ground beetles may be anomalous. Niemela (1993) demonstrated that they show a bimodal abundance distribution, with the few abundant species and more numerous rare species not linked by a continuum of intermediately common taxa. Several possible explanations for this anomaly were advanced, one being that only a few species had adapted sufficiently well to the boreal forest to be able to become abundant.

For much of the tropics, the natural abundance of many beetle species is unknown. Floren and Linsenmaier (2003) noted 'most Coleoptera of primary forests are extremely rare and faunal overlap in samples is very low'. Incidence of many low-abundance species is endorsed by numerous faunal studies in the tropics. In Papua New Guinea, Allison *et al.* (1997) collected 418 beetle morphospecies (in 53 families) by fogging eight individual trees of *Castanopsis acuminatissima*, and 199 of these were represented only by singletons. A further 83 morphospecies were each represented by two beetles, so that a high proportion of the taxa was apparently rare. Similar trends have been noted elsewhere; for example, Stork's (1991) 859 beetle species from canopy fogging in Borneo included 499 represented by singletons and a further 133 with two individuals. And a classic Amazonian forest beetle study (Didham *et al*. 1998a,b; see p. 95) yielded 45% of singletons across 993 morphospecies. Such patterns of relative abundance seem to be general. The *Castanopsis* study above exemplifies anther relevant facet of beetle diversity, namely that single plant species, or even parts, may support substantial diversity, perhaps as a specific critical resource on which some of those species depend. As another example, woody petioles of *Cecropia* trees (four species) in Costa Rica yielded 36 species of beetles, most of them Scolytinae or zygopine weevils (Jordal & Kirkendall 1998).

Tropical beetle richness, even on single tree species, can indeed be impressively high. However, increased sampling effort also detects changes over time, not simply through increasing take of what is already there at any given time. For Galapagos beetles, for example, Peck (2006) noted that the continuing dynamic pattern of faunal change, resulting in part from human activity, can be revealed only by continuing investigation. The number of species known there from only one or two specimens suggested the likelihood of others being present, but Peck also noted that poorly investigated habitats and novel sampling methods to augment earlier capability should be conducted routinely in such attempts to augment inventories. There will inevitably be some form of trade-off between sampling effort and the time/resources available. Most commonly, cost and time available will not permit surveys of indefinite length and complexity, and very careful planning is needed to optimize the field procedure in relation to answering precise questions and accepting the compromises that ensue. For dung beetles, Davis (2002) noted one such compromise as being failure to collect some of the rarer species at each individual site whilst increasing the number of sites sampled in order to increase the geographical representation of a survey. If assessing sites for typicalness or representativeness, determining the consistently present members of a beetle assemblage may be more relevant than retrieving every very scarce species present. Their presence may, of course, add conservation significance to a site but this may not be the primary aim of the programme. Again for dung beetles, Hanski (1982) differentiated between core and satellite species in assemblages, a principle of very wide relevance in clarifying community structure. The core species are those that are relatively common and present in all or most suitable sites all the time, not necessarily as ecological specialists but as reliable elements that can help to characterize the assemblage. Satellite species, in contrast, are rare (or more sporadic in incidence) and may frequently become extinct and be sustained by repeated establishment of new populations at sites, perhaps as metapopulations so that knowledge of regional dynamics becomes a central theme in their evaluation. Scale effects may be important in assessing this aspect of assemblage dynamics (see p. 65). For most groups of beetles, however, nothing is known of population dynamics or the factors that cause many species to be as rare as they appear to be. And, despite the academic attraction of undertaking more comprehensive surveys, short surveys may prove perfectly adequate for many purposes.

The presence or absence of threatened species, either particular individual species or broader representation from a local directory such as a red list or red data book, is widely viewed as fundamental information in designing and implementing conservation. However, the amount of sampling needed to detect all or even most such species is difficult to define. For boreal forest beetles in Finland, numerous saproxylic beetles, in particular, are of concern as regionally extinct, threatened or near-threatened (Table 1.4). The beetles of these forests have been studied intensively over several decades so that the conservation status of many species is reasonably unambiguous, and the presence of rare or threatened species in samples is used to indicate the conservation values of individual forests or to help dictate sympathetic forestry management. Martikainen and Kouki (2003) attempted to address the problem of the sampling effort needed to assess the presence of the significant beetle species, using window traps attached to the trees. The overall number of beetle species trapped in an area was a useful indicator of representativeness, and an almost exponential relationship was found between numbers of total beetle species and conservation interest species. Samples comprising fewer than 200 trapped species or 2000 individuals were considered 'almost useless' in surveying threatened or near-threatened species, and the probability of finding these taxa increased markedly when the number of beetle species trapped was greater than 400. In essence, very large sample sizes were needed, even using well-understood and effective sampling methods in order to reliably rank different forest patches for significance based

Table 1.4 Forest beetles in Finland: numbers of regionally extinct, threatened (divided into critically endangered, endangered and vulnerable) and near-threatened species, indicating the assemblage importance and diversity of saproxylic beetles.

| | Regionally extinct | Threatened | | | Near-threatened | Total |
		Critically endangered	Endangered	Vulnerable		
Saproxylic species	12	33	48	56	65	214
Other species	9	6	13	27	23	78
Total	21	39	61	83	88	292

Source: Martikainen & Kouki (2003) with permission.

on diversity of such notable beetles, in a fauna that can be regarded as well documented. In their example, Martikainen and Kouki (2003) noted that ranking 10 boreal forest areas in this way may require trapping of more than 100,000 individual beetles, analysis of which is simply not feasible as a routine exercise.

The presence of such notable species in samples is commonly largely a matter of chance, with their rarity (as low abundance and restricted distributions) rendering any attempt to detect them uncertain. Even enumerating the species not trapped is highly uncertain (Muona 1999) and, if possible at all, relies on extrapolation from previous records probably resulting from different methods, possibly long ago and in different (pre-disturbance) environments, and even from misidentifications. Sampling interpretation for most tropical forest habitats is far more difficult than for Finland, because of the larger overall numbers of beetle species present and because many or most will not be described or named, so that their conservation status will not be definable easily. Apparent rarity is not necessarily equivalent to vulnerability.

For Iberian water beetles, Ribera (2000) recognized four categories of rarity, and these have much wider relevance in conservation assessment, as they can infer very different conditions and security.

1 Species which may indeed have greater distribution and/or abundance than known, for example recently discovered species whose extent has not been explored beyond, possibly, the single site or population from which they are at present known.
2 Relict species, those which are now the isolated remnants of formerly wider distributions, perhaps as a result of long-term ecological change, such as in some alpine species.
3 Rare local endemic species, perhaps restricted to very small and specific sites (in Ribera's example to selected permanent stretches or headwaters of individual rivers or streams), and for which site protection is needed.
4 Species that have demonstrably become rarer as a consequence of human activity, as threatened species, for which the causes of loss or decline can sometimes be unambiguous.

However, such approaches are also predicated on adequate definition of which beetle species are acknowledged to have conservation interest or significance, rather than simply being rare or elusive. The term 'rarity' has a variety of meanings, as used above, but implies some form of scarcity and, perhaps, predisposion to threat. Although rarity and threat are commonly compounded or confused in conservation assessment, vast numbers of insect species are naturally rare but not necessarily threatened. The distinction is often unclear, simply because putative threats cannot be evaluated easily, and the three conditions of rarity (namely low abundance, narrow distributions and ecological specialization; Rabinowitz *et al.* 1986) may predispose a species or population to stochastic influences or localized threat. Although it is common for only one of the above three states of rarity to occur in a species, any combination of them can occur. The vulnerability of the tiger beetle *Cicindela deserticoloides* in Spain, for example, results from low abundance, small geographical range and habitat specialization (Diogo *et al.* 1999), so embracing the three parameters.

Evaluating conservation status and significance

The most widely accepted rules for evaluating threat to individual species flow from systems developed through the World Conservation Union (IUCN 1994, 2001), which involve assessment of each species against a suite of criteria to evaluate risk of extinction. These categories are illustrated in Fig. 1.3 and are the basis for placing species formally on a red list or similar document.

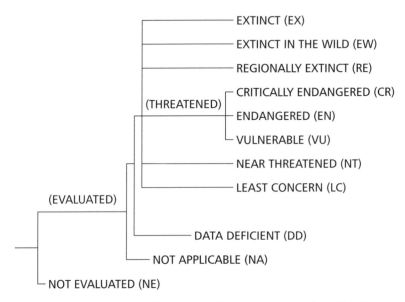

Fig. 1.3 Schematic summary of the IUCN Red List Categories, in which Threatened includes the three categories of Critically Endangered, Endangered and Vulnerable. (From IUCN 2001 with permission.)

Nevertheless, applying the IUCN Red List categories reliably to beetles is difficult; in common with almost all other insects, we usually have no data on population sizes and numerical trends of decline, or of the factors that enable estimates of probability of extinction. In almost every case, only selected criteria may be able to be applied. Most such concerns over threat have arisen from demonstrated loss or change to habitats, and thus related to declines in extent of occurrence or area of occupancy. Often, this is the only reasonably reliable information available and can transcend national or other political boundaries. For the 30 red-listed species of Cerambycidae in Finland (of a total of 81 species), Komonen (2007) classified the species variously as resource-limited (host plant), substrate-limited (to particular host plant attributes), or climate-limited. In Finland, many of the red-listed species are represented by peripheral populations on the edge of a wider European range, so that their importance nationally may be far greater than that over their full geographical range. Factors that limit distribution of beetle species, and which may create localized distributional ranges that accord them conservation interest, have frequently been suggested. For most, the historical events leading to the present distribution are not sufficiently known so that the twin scenarios of local endemism and relictualism (the latter reflecting survival after disappearance elsewhere as a consequence of habitat loss or other threats; see p. 72) may become confounded. Nevertheless, as indicated in particular by several studies of scarab beetles, the factors associated with extent of environmental tolerance or specialization and with dispersal capacity, are both highly relevant. Dung beetle distributions are thereby influenced by the amount and kind of dung available, reflecting in turn changes in the native mammal fauna or grazing stock regimes. Allsopp (1999) speculated that there might even once have been some Australian dung beetles associated with the prehistoric mammalian megafauna, extinct since the end of the Pleistocene. In Spain, limited distributions of *Jekelius* species are influenced substantially by the acidic or basic nature of the substrate, in conjunction with limited dispersal potential of these flightless beetles (Lobo *et al*. 2006).

Successive global red lists of threatened animal species each include several hundred beetle species allocated, sometimes tentatively, to one or other category of threat or regarded as 'near threatened' or 'data deficient'. The listed taxa are geographically widespread, but actual numbers are sometimes uncertain. In addition to series of individually named beetle species, earlier directories in this series include entities such as 'all species of genus X'. The most recent version of the *IUCN Red List of Threatened Species* (IUCN 2008) includes only 72 individual extinct or threatened beetle species, reflecting revisions from earlier lists, and uncertainties of current status. The major message, though, is that most beetles have not been evaluated on this scale, and the magnitude of conservation need based on numerical or range declines or habitat losses is likely to be far greater than appreciated widely at present. More complete directories are available for many countries, particularly in Europe, and indicate much higher levels of loss for some national or regional beetle faunas. Thus the UK Biodiversity Action Plan, designed in 1999, listed 54 beetles among its designated priority species, and many of these have received considerable attention to clarify their conservation status and needs. Later this number was increased to 87 species, but recent reports suggest that four of these have become extinct and that the future of others is parlous. More widely, around 250 of the UK's 4000 beetle species have not been found in the wild since 1970, and their fate is largely unknown. In contrast, more local scales of conservation reflect more local losses, so that conservation concerns arising from local threats, or to species perhaps more abundant elsewhere, are important. One British example, of many that could be cited, is for the flightless bloody-nosed beetle (*Timarcha tenebricosa*) in Warwickshire. It was formerly widespread in the county, but is now known from only two relict populations on small sites. Imminent threats include site losses through airport development and road construction. A local action plan (Lane 2003) sets out a number of measures designed to safeguard the beetle.

Any list of designated priority species is a useful initial filter for demonstrating conservation need, and collective benefit may be increased if ecological variety is represented among the taxa. For Britain's *Red Data Book*, Shirt (1987) listed 142 endangered, 84 vulnerable and 266 rare beetle species, collectively around 14% of the British fauna, a level which, although daunting, is not surprising. The most important habitat associations represented, by number of dependent beetles, include woodlands (40%, particularly ancient woodlands with about 90 endangered or vulnerable beetle species), coastal situations (21%), wetlands (19%) and grasslands (11%). For such well-documented faunas, it is possible to review the status and conservation significance of a very high proportion of the species present, and the review by Hyman and Parsons (1992, 1994) illustrates the kinds of information that can be derived from biological knowledge combined with distributional data from recording schemes (below). These form the foundation of many conservation exercises and plans, and aid considerably in establishing priorities among species. Many of the species are known from single sites or populations, others from very few localities, and each may independently merit and need urgent conservation measures. However, as Haslett (1997) put it, in any such listing or compendium of invertebrates we are 'spoiled for choice' as there are simply too many species that qualify for inclusion. Haslett advocated some focus in enlarging such lists to include representatives of species associated with habitats that are under-represented, rather than simply adding 'more of the same', so that those species could be representatives of ecologically functional groups essential to the continuity of the ecosystems concerned. Some such species (such as *Cerambyx cerdo*; see p. 45) and the cetoniine *Liocola lugubris* in Europe assume the roles of keystones whose well-being reflects that of numerous other coexisting taxa. In this case, widespread declines of *L. lugubris* may influence many other saproxylic species, and protection of ancient deciduous woodlands in part reflects the conservation needs of this beetle. Many localized or ecologically specialized beetles can be promoted as symbols or flagships for particular habitats or sites, with striking appearance or unusual or novel biological characteristics increasing public interest. Particularly notable species, such as the European stag beetle *Lucanus cervus*, can be important promoters of interest in insect conservation (Smith 2003). Individual species conservation plans (below) may be important in drawing attention to unusual habitats or sites. Thus, the North American delta green ground beetle (*Elaphrus viridis*) is one of a varied suite of taxa that highlights the significance of vernal pool ecosystems in California. Species plans may be complemented effectively by those focusing on suites of species, related either taxonomically or ecologically: British plans include those for three such ecological groupings of beetles, namely river shingle beetles, *Harpalus* spp. and saproxylic beetles, whereby suites of species with biological features and conservation needs in common can be appraised together (see p. 199). As a North American example, three beetles are among the total of seven invertebrates that highlight the importance of karst caves in Texas. This important theme is discussed further on p. 199, and is important in cases (i) where such species co-occur and are not sufficiently well understood to enable individual conservation plans to be prepared, or (ii) where the major need devolves on habitat protection for all the species involved so that conjoint effort is efficient.

Placing species on lists accords them priority for conservation, but the lists themselves vary considerably in primary purpose. Some are advisory in not carrying legal weight, whilst others are more formal documents that oblige action and responsibility. It is not uncommon for initial advisory listings to form the basis of regulations at some future time. Listing is thereby a responsible step, likely to influence how a recorded species may be treated in the future (New 2007) and it may take some time to consider and approve a nomination. At the least, a listed species may gain notoriety and additional publicity. However, formal listing is often accompanied by some form of prohibition of take, as a perceived threat to the species. For collectable organisms, such as many beetles, this prohibition may lead to illegal black market trade with high prices offered and paid for threatened species, including international poaching and smuggling and sometimes involving substantial damage to sensitive habitats by unscrupulous gatherers. Such provisos seem to apply to the several Lucanidae protected formally in Taiwan for example. Several examples are discussed later, but a subsidiary effect of protecting threatened (or putatively threatened) species in this manner is that hobbyists may be frustrated by attendant needs for permits and have their interest discouraged by an atmosphere of suspicion, causing them to transfer their leisure time to other pursuits. It is better known for butterflies (see Greenslade 1999) that the very people whose continuing interest is the major channel through which greater understanding of rare species can be accumulated and incorporated into informed conservation planning can be alienated by ill-planned prohibitions. Sometimes, these are not seen to aid conservation and alone are unlikely to reduce or remove threat. Decisions to designate species as protected must be responsible, transparent and justified after canvassing the widest possible advice from people who understand the taxa involved from field experience.

Further problems can arise when calls are made to formally protect beetles (or other insects) that are members of groups containing many very similar-looking species, simply because the individual species may be recognizable only by a specialist, because the insects are small or because the specific characters need careful appraisal, perhaps involving dissection or measurements. In some groups of beetles, individual insects of the same species may differ considerably in appearance or size. Indeed, one basis of desirability of beetles to collectors involves these features, with large or heavily ornate individuals of scarabs or stag beetles sometimes commanding far higher prices than smaller or less ornamented individuals of the same species. Most people (such as customs officers) responsible for enforcing any prohibition of take or trade are not specialists in the taxonomy of beetles or even entomologists or biologists, and it is both unreasonable and impracticable to expect them to recognize most individual species within these groups. Likewise, most ecologists examining assemblages or large multispecies samples of beetles are highly unlikely to initially recognize particular protected species of small beetles in their voucher series.

One avenue towards overcoming such problems is to widen the ambit of formal protection. Thus, listings of 'all species of genus X', noted earlier, may serve the purpose of protecting individual included species that merit this by including them with their close relatives with which they might be confused easily. Notwithstanding that some of those relatives may be abundant and of no current conservation significance, their protection acts as an umbrella for protecting the truly needy species. The approach can be contentious, but is simply a manifestation of the precautionary principle. In a few cases, such steps have involved a collectable group of beetles. The formal listing for protection of all species of jewel beetles in Western Australia by the state government in 1978 was motivated largely by their stated desire to protect some rare collectable species from trade, but caused considerable disquiet among hobbyists. The credibility of this legislation for conservation was thrown into doubt by subsequent authorized destruction of substantial areas of prime jewel beetle habitat within the state: protective legislation is not in itself conservation, but such listings may sometimes be an effective first step to enable conservation to occur.

Collins (1987) provided a list of specific Coleoptera selected for formal protection in many parts of Europe, either nationally or in particular regions within a country. A greater array of countries listed Lepidoptera, and in almost all places the list of beetles was considerably shorter than that for butterflies and moths. Indeed, many countries listed less than a handful of beetles, occasionally only one. However, as above, some broadening occurred sporadically. Thus, in Salzburg (as one of the separate Länder of Austria) the protected taxa list included 'all species of Cerambycidae except *Hylotrupes bajulus*', and that for another region 'Scarabaeidae: Cetoniinae'.

Particular beetles have long been heralded as of conservation concern but, as noted above, the real scale of the problem they face is vastly underestimated by lists of such species. Wells *et al.* (1983) included seven beetles in the first *IUCN Invertebrate Red Data Book*, sufficient to indicate the variety involved, but these are simply examples of the numerous species meriting such concerns, for a variety of different reasons. Lists of signalled species are invaluable initial guides to the status of regional faunas or taxonomic groups, but for such poorly known animals as beetles they reveal two main categories of problem. First, lists are almost invariably too short to be fully representative, and simply reveal the tip of the iceberg of needy species that have received sufficient attention to be nominated and validated for inclusion. Second, even though the lists are so short, they commonly still include far more species than can be properly managed or can receive adequate individual treatment from the resources available for practical conservation. Whereas single-species studies and management remain the most tangible and popular level of beetle conservation to many people, many other beetles are necessarily conserved only under the umbrella of wider studies. Nevertheless, many individual beetle species clearly need practical conservation, and debate will continue over how the most deserving targets are best selected, within the widely accepted framework that the most needy species may be accorded priority. However, basic biological knowledge of many threatened species is very poor and their scarcity renders them difficult to study effectively in order to improve this situation. There may be considerable uncertainty over where they occur, the form and size of their populations, factors causing conservation concern and even whether the species is still extant.

Recording schemes for beetles are proliferating in efforts to more accurately assess species' distributions and abundance, to help detect trends of decline or actual losses, and to generally improve the level of basic knowledge needed to properly assess conservation status and needs. Recording particular species, of course, depends on our ability to detect and recognize that species. Even for some spectacular and nominally well-understood beetles, difficulties can arise. The European stag beetle *Lucanus cervus* is a notable flagship species but, notwithstanding that around 1300 recorders participated in a survey of its distribution in Britain (Smith 2003), distributional data for some other parts of it range are still sparse. Thomaes *et al.* (2008) suggested that many sites remain undetected, reflecting that *L. cervus* has a very short adult flight season and is nocturnal, so that it is substantially under-recorded to the extent that data for reliable designation of suitable protected areas are not available. Particularly in parts of Europe, recording schemes can draw on substantial accumulated knowledge, and any such endeavour can include two main sources of information.

1   compilation of published records, and the transcribed label data from specimens in museums and private collections, are sources of historical information that may span a century and more;
2   current surveys, in which the most recent systematic arrangement can be applied to target particular habitats or taxonomic groups.

Other than for easily identifiable (generally equating to collectable) groups, anomalies of species naming are likely to persist through the historical record, so that the existence of voucher specimens to validate identifications is of critical importance. Categorizing accumulated records by time interval can indicate possible changes in abundance and distribution, but the data are almost inevitably sporadic and patchy, so that considerable care is needed in extrapolation. Other than for more systematic or comprehensive recording, such as occasionally for particular reserves in Europe, changes in abundance are almost impossible to assess from this information. The growing number of databases, of ever-increasing sophistication and incorporating reliably identified museum records and current data, are an invaluable investment in assessing changes in the future, as indeed are the specimens themselves. Unidentified material from ecological surveys, archived as 'ecological collections' (Danks *et al.* 1987), may be of critical importance as human demands on land and water proliferate. Even for better-documented insect groups such as butterflies and dragonflies, increased numbers and completeness of entomological surveys in recent decades may yield results far more complete than earlier records based on the efforts of a few recorders or enthusiasts. Many early recording schemes thus reliably include presences while perhaps including data from only part of the range, but sampling effort may not have been comprehensive and the interpretation of purported absences may be difficult.

Such databases are the template for assessing trends and change, based on species incidence, and can help to determine changes that have already occurred and as models to predict those anticipated in the future, for example as a consequence of climate changes (see p. 133). Beetle distribution recording schemes, in Britain and elsewhere in Europe in particular, are providing much information of conservation relevance, not necessarily restricted by political boundaries (see p. 39). The most thoroughly appraised example for insect recording is for the British butterflies, based on more than a century of recording a small and well-studied fauna, mapped on a base scale of 10×10 km squares (see Asher *et al.* 2001), and admired as a model for emulation elsewhere in both methodology and detail. This has allowed convincing interpretation of range changes in British butterflies and has become a major foundation for conservation activity and planning. Recording schemes for beetles have not yet achieved such venerability, but many are indeed accumulating, most focusing on beetles of particular families or habitat associations. A 10×10 km square represents a huge area to a beetle and each such unit is likely to include numerous different habitats. Nevertheless, as Eyre *et al.* (2006) noted for water beetles in Britain, 'there is little doubt that 10-km square records do constitute a measure of variation at the large biogeographical scale'. Data on even a few species can be revealing. With selected examples from only two families (Cantharidae, Buprestidae, drawing from a recording scheme for these groups started in 1984) in Great Britain, Alexander (2003) illustrated species that have remained largely stable in distribution, expanded their range, or declined considerably. The last of these give valuable clues to conservation need, not least because many declines can be linked with particular facets of habitat change attributed directly to human activity. Thus, declines were found in areas associated with agricultural intensification, and among beetles associated with ancient trees (see p. 84), coppice woodland and open woodland affected by changing management practices. Other range changes may be linked with climate changes (see p. 133).

One such example from elsewhere is for dung-rolling Scarabaeidae in Italy (Carpaneto *et al.* 2007), in which declines of these species were appraised using data accumulated since the 19th century and categorized by decades in seeking possible trends. The data included all literature records from 1865 to 2004 and 1413 unpublished records from collections, from all 20 administrative regions of Italy [within which 282 UTM (universal transverse Mercator) grid cells with more than 15% land area were assessed separately for records], to give a total of 6870 individual records. Three patterns of decline were suggested by frequency of records, as species starting to decline in the 1960s (two species), 1970s (three species) and 1980s (six species) respectively, so that all 11 species in the fauna manifested apparent declines. Several species appeared to have disappeared entirely from northern regions, and six were considered to have a high risk of extinction nationally. Even allowing for considerably greater recording effort in more recent times, these trends appeared real. Declines were attributed in part to changes in livestock systems, from predominantly free-ranging cattle to stabled stock with consequent unavailability of dung in the field, linked in part with loss of open pastoral areas to forestry or intensive agriculture. Increased predation on beetles by hooded crows (*Corvus corone cornix*) might also be a contributing threat.

Records of species incidence, whatever method was used to obtain them, can (once accepted as valid) be used to map distributions, and a number of beetle atlases for Britain or parts of western Europe are important aids to help demonstrate conservation status and its changes in individual species. In particular, declines of species may be linked with particular factors in the area, most commonly habitat change or loss. These are the only parts of the world where accumulated records of beetles identifiable to the species level across major parts of faunas have been made for more than a few decades, so that changes in incidence and distribution detected by comparing maps made for different periods may be based in reality, rather than simply reflecting sampling unevenness. Mapping or recording schemes may commonly arise from non-systematic sampling (Lobo *et al*. 2007, on Iberian Scarabaeidae), which later become more comprehensive as their values are recognized and consolidated, although historical biases over knowledge of species' distributions are likely to remain widespread for any group.

One notable example (Desender & Turin 1989) was developed from records on 419 ground beetle species recorded in Denmark, the Netherlands, Belgium and Luxembourg, and published in earlier atlases of Carabidae from these countries, each enabling comparison of the fauna before and since 1950. A total of 281 species were recorded in all areas, with Belgium and Luxembourg treated together. Most of the other 138 species recorded from one or two areas are rare, so that any estimates of their decline are not necessarily reliable. Nevertheless, many were regarded as 'seriously threatened', by decreased extent of occurrence within either the main section of the area (nine species) or the whole area (eight species), or 'threatened' with implications of wider decreases (37 species). Another 11 species were noted as 'probably threatened'. Of the wide-range species from all three areas, trends were appraised against ecological attributes (stenotopic to very eurytopic and tolerant to cultivation), geographical range in relation to being centred in the Netherlands, and habitat affinity on an 8-point categorization of xerophilous, more-or-less xerophilous, mesophilic or cosmopolitan, hygrophilous, more-or-less hygrophilous forest species; preference for shady sites and bushes; arboricolous; and synanthropic. Collectively, 142 species were considered endangered, with many apparently having disappeared. Declines were particularly high among xerophilous species, many of the stenotopic species and those found in habitats such as heathland and low-quality grassland. Drawing from the Netherlands data of that survey, Turin and den Boer (1988) considered that progressive isolation and loss of suitable habitat fragments may be a component of decline, as many of the carabids lost were poor dispersers. They emphasized the substantial loss of Netherlands dry heathlands from around 800,000 ha in 1835 to only about 40,000 ha by 1980, with additional degradation of much of the remainder by grass invasions. This scenario was revisited by Kotze and O'Hara (2003), who showed that the carabids that have declined are commonly the larger-bodied species and habitat specialists. However, flight dimorphic species (see p. 176) had been less prone to declines than species that were either wholly flight capable or wholly flightless, possible reflecting additional benefits of the ecological 'bet-hedging' that accompanies the twin strategies of dispersal capability and obligatorily staying put. Nevertheless, problems remain over clarifying fully the reasons for these declines. However, without the temporal base afforded by the recording schemes, no such sound template for discussion would be available.

Related applications include that among the 10×10 km recording units being used to classify these areas by the ground beetles present (Hengeveld & Hogeweg 1979), nine groups of squares were distinguished on the carabid representations. Estimates of land cover to correlate with beetles were broad, but clearly demonstrated the potential to use beetles in this context.

In another important contribution to understanding distributions of beetles in landscapes, Eyre *et al.* (2003a) used carabid data from Britain (namely presence/absence data of each of 356 species from 1687 recording squares) combined with land cover data derived from satellite information in order to determine the extent to which land cover may be able to predict the ground beetle species pool. Nine groupings of beetles were detected: three showed strong relationships with upland ground cover; three others were associated with deciduous woodland, coastal and tilled land; and three others, although not associated strongly with any particular form of land cover, differed in geographical position. However, because many of the cover variables were closely associated, it was sometimes unclear which factors the beetles were responding to. Eyre *et al.* suggested that, with further analysis, the relationships between carabids and land cover might indeed lead to their wider use in monitoring environmental changes across the countries. By analogy with the Netherlands data, this might extend even more widely across Europe.

2

Practical Conservation: Basic Approaches and Considerations

As for any species or assemblage, practical conservation of beetles draws on a considerable variety of ecological themes. For example, the distribution of many species is restricted and thus is a major contributor to their perceived conservation need; as a result, conservation attention may focus strongly on particular sites within a formerly broader distributional range. Site security is then the most important initial step for conservation: simply, without a place to live, those beetles cannot persist in the wild. In the past, however, securing key sites (such as by including them in reserves) has sometimes been viewed as all that is necessary for conservation. In reality, it is usually only the first (but essential) step in facilitating continuing management to cater for the needs of the species or assemblage within that area. The various measures will be guided by informed knowledge, often necessarily augmented by original targeted research, of resource needs and actual and/or potential threats as the twin themes that underpin practical insect conservation. Levels of knowledge are almost inevitably inadequate for formulating ideal management at the outset.

The two major categories of conservation-related study involving beetles encompass the need to (i) conserve nominated or selected beetle species in particular habitats or biomes or on particular sites, and (ii) sustain or restore the condition and integrity of those biomes as signalled by the spectrum and abundance of beetles present. For all such studies, a widespread interpretative problem arises from determining what is optimal for a species, or in adequately characterizing the species composition of assemblages that may be desirable. Using Hanski's (1982) core–satellite species concept (see p. 32), Niemela and Spence (1994) examined the distribution of Carabidae in temperate deciduous forests in North America at scales ranging from local to continental, in relation to environmental

Beetles in Conservation, 1st edition. By T.R. New. Published 2010 by Blackwell Publishing.

variables. At the smallest (local) scales, factors such as tree cover, understorey cover and occurrence of other carabid species were correlated with distribution of particular species. Distribution and abundance of many species were positively correlated, and distributions of 10 core species aggregated at all scales. All 114 species in the continental dataset were correlated in this way. The major conservation inference is that carabid distribution is determined by different factors at different ecological scales, with the effects of those factors not necessarily transcending scales. Core species tend to be most often habitat generalists, in contrast to more restricted satellite species. The particular resources or climate factors influencing heterogeneous distributions and localized associations are commonly not known.

As with other insects, it cannot be presumed that other studies, for example on closely related species or even on the same species in a different environment, can be transferred uncritically and wholly to the new situation. Although these may indeed form a strong foundation, many variables occur, such as seasonality or life cycle, food spectrum, the local array of competing species or natural enemies, and localized or particular threats (together with capability to counter their effects). Such differences have resulted in much species-level conservation of localized insects being largely site-specific, with management of each site drawing from (and in turn contributing to) the pool of relevant knowledge and experience. In many cases, conservation need is demonstrated only retrospectively. Surveys of sites or protected areas can unexpectedly reveal unusual or threatened species or associations. Should those species be listed, some formal conservation investigation or protection may be triggered by regulation. In other cases, the enthusiasm or interest of the discoverer or site/area manager may be enlisted to promote conservation interest. Many beetles are among the beneficiaries of such fortuitous discovery and its consequences.

Beetles are thus participants in conservation as focal targets, whereby the species or assemblage is the object for conservation, or as tools that can provide information on site or broader environmental quality and so aid in ranking or categorizing areas or systems for conservation priority. Reflecting the variety of beetles likely to be present on a ground area or within a freshwater body, site value assessments can draw extensively on data on beetle richness and abundance. In Britain, and progressively elsewhere, the importance of national or regional recording schemes includes development of indices of site quality, based on rarity scores to help rank sites. Both water beetles (following Foster 1987; Foster & Eyre 1992) and Carabidae (following Eyre *et al.* 1996, 2001) have been used to demonstrate the principles and potential of this use. Carabidae in parts of Scotland and northern England (Eyre & Luff 2002) exemplify the kind of information that can be applied. Site quality score (SQS) was based on species rarity score (SRS), with the latter derived from beetle atlas data at the 10×10 km scale (see p. 42). For Scottish beetles, each species was given a value based on a geometric scale (1, 2, 4, 8, 16, 32, 64) depending on the number of such squares in the country atlas in which it had been recorded since 1970 (numbers of squares in sequence for each of the above categories 1, 2–3, 4–7, 8–15, 16–31, 32–63, 64 or more). For northern England, the more complete data allowed the post-1970 data to be used at the finer 2×2 km scale The SRSs taken from all species at a site are summed, and divided by the number of species to give the SQS, which can be used to compare rarity value of sites grouped within habitat classifications. Likewise, the measure of fidelity noted for exposed riverine sediments (see p. 180) can also draw on recording schemes for refining beyond the simple division noted there.

The principle of such indices is to delineate areas (usually specific sites) of particular or elevated importance for conservation. In the relatively few cases where distribution of species is well mapped within a greater area, the procedures demonstrated by Foster (1987) for water beetles are invaluable for this purpose. However, most beetles are not as well mapped, and only less precise species scoring methods thus apply. As another example, the saproxylic quality index (SQI) described by Fowles *et al.* (1999) drew on a wider British recording categorization based on common species (score 1), local and unknown species (score 2), regionally notable species (score 4), notable/notable B species (score 8), notable A species or Red Data Book K species (score 16) and Red Data Book 1–3 species (score 32) (see Ball 1985 for explanation of these categories). Naturalized species are ignored in this scheme. The SQI is obtained by summing scores for all species with individual scores of 2 or more and dividing this total by the number of saproxylic species present, including those scored 1 and which did not contribute to the index total. Successful use of any such scoring system depends on availability of complete or near-complete lists of species present.

Species importance

Species-level conservation remains the most familiar and tangible level of practical conservation to many practitioners, but setting priority among the overwhelming number of potentially deserving candidates remains problematical. It commonly involves some form of triage, not necessarily confined to intensity of threat, to select the species most deserving of the limited conservation effort available. A range of ethical issues, biological evaluations and estimates of extent of threat all contribute to this selection together, in some cases, with the portfolio of candidates being formally delimited by a list of designated threatened species eligible for agency attention. Practical ecological importance of individual species is often difficult to demonstrate, but there is little doubt that conventional flagship or putative umbrella roles commonly advanced for species with little direct evidence of ecological interactions may sometimes be enhanced. Additional values of beetles as ecological indicators may become strong motivators for their well-being. Within the saproxylic insect guild, for example, suggestions have been made that particular species may facilitate the well-being of assemblages through a role as ecosystem engineers.

The striking European longicorn *Cerambyx cerdo*, one of the largest Cerambycidae in the region with a body length of around 55 mm, breeds in oak trunks, and is one of the species targeted for conservation under the European Habitats Directive for maintenance of existing populations and long-term survival. It is protected as an endangered species. Populations have declined largely as a result of loss of oaks and oak-dominated landscapes by agricultural and forestry management, and the formerly widespread beetle has been reduced to a few isolated populations in some countries. It is extinct in the UK and Sweden but has become a flagship species among saproxylic beetles and has received considerable conservation attention, not least because it has been claimed as a possible ecosystem engineer (Buse *et al.* 2008) through altering its own habitat to create favourable conditions for a variety of other threatened beetles. Its presence and abundance may thereby facilitate well-being of entire assemblages of saproxylic insects. This inference flowed from comparison of beetles (taken by flight interception traps) on oaks colonized or not colonized by *C. cerdo*, showing that colonized trees supported considerably richer assemblages. Physiological differences between the two categories of trees suggested that these might also be linked with *C. cerdo*. The alternative, of physiological differences being the attractant principle for beetle attack, was considered less likely. Beetle species richness was correlated more strongly with intensity of *C. cerdo* colonization than with either sun exposure or trunk diameter, so that presence of the longicorn is regarded as a useful indicator of saproxylic beetle richness. Based on this engineering role, Buse *et al.* suggested that reintroduction of *C. cerdo* might be considered to regions where it had become extinct, if a sufficient number of old oaks are available, with some desirable habitat parameters discussed by Buse *et al.* (2007).

A somewhat similar role has been attributed to another cerambycid, the large Neotropical harlequin beetle *Acrocinus longimanus* (in which the males have dramatically elongated forelegs), which is considered a pioneer species in the foundation of saproxylic invertebrate communities in dead and decaying trees of the Moraceae and Apocynaceae (Zeh *et al.* 2003). Larvae bore into the wood and the accumulated sawdust creates the substrate colonized by a variety of other invertebrates. The beetle was categorized by Zeh *et al.* as a keystone species for such communities.

Equivalent roles have only infrequently been investigated in conservation contexts, but should not be unexpected among such an enormous and varied group of organisms, by which beetles form rich functional feeding guilds in a wide variety of biomes. However, in some ecosystems their ecological role may indeed be rather small. In an appraisal of arthropods as 'webmasters' in desert ecosystems, Whitford (2000) considered only termites and ants as keystones for example. Ecological influences of beetles may be predominantly on and above the ground, other than those of some Scarabaeidae.

Nevertheless, the ecological impacts of beetles are demonstrated dramatically by some abundant species, with those of some beetles involving landscape-changing effects as a powerful manifestation of their activities. The southern pine beetle *Dendroctonus frontalis* (Curculionidae: Scolytinae) exemplifies the effects of such herbivory. This bark beetle, one of several important pest species of *Dendroctonus* in North America (see p. 3), has three major target groups in conifer forests (Coulson & Wunneburger 2000): (i) acceptable host groups, involving several commercially important species of *Pinus*; (ii) susceptible habitat patches, stands that may occupy up to several hectares and with mature trees that can act as nuclei of infestation; and (iii) lightning-struck trees. In the last, the resin defence system is apparently broken down, so that the trees become highly attractive to beetles and act as centres for initiation of potentially wider infestations, refuges for dispersing beetles, and stepping stones linking beetle populations between other habitat patches. Coulson and Wunneburger categorized *D. frontalis* as an 'agent of creative destruction', by which forest landscape structure is affected through tree mortality from beetle herbivory.

Many workers have proclaimed the value of beetles as indicators, in some way effective surrogates of wider biodiversity in responding to environmental changes, and so providing information on their effects. Extensive studies have sought to demonstrate, for example, the changing incidence of beetles in modified ecosystems such as along natural–anthropogenic gradients, with the twin highly altered arenas of agricultural and urban ecosystems the most frequently assessed. Changes in beetle faunas have been explored in many aspects of landscape management, such as restoration and rehabilitation. Any study of surrogate taxa must have very clear purpose, and assess what might be gained or lost by approximations, whether reduced taxonomic penetration or selection of some groups to represent others. In a recent survey of biodiversity assessment in agricultural ecosystems, levels assessed included (i) number of arthropod orders, (ii) number of families of Coleoptera and (iii) number of species of Carabidae, based on pitfall trap catches in two different systems. Order-level analysis (26 orders), on the face of it overly simplistic in using very broad categories (which, however, are easily identified by non-specialists and thereby attractive in rapid assessments), gave discriminations for land-use equivalent to those of carabid species (23 species). Family-level analysis of beetles (50 families) did not show any parallel distinctions (Biaggini *et al.* 2007). In these samples, four species of Carabidae collectively comprised almost 84% of the total of 890 ground beetles, so that trend differences reflected incidence of numerous rare species and variable representation of a few common ones.

In some groups, parallels occur between ecological variety, diversity and potential indicator value, as reflected in their varying responses to any given change imposed on an environment. Thus, Bohac (1999) emphasized ecological variety within Staphylinidae, aided in interpretation by the easily recognizable and diagnostic appearance of these abundant and diverse beetles. The five major groups of life forms of Staphylinidae (Table 2.1) are each further divisible into lower consistently recognizable groups, and their representation can differ substantially in number and proportion in particular biotopes. Bohac indicated the trends with increased habitat change as including (i) increased proportion of generalist (eurytopic) species, (ii) increased proportion of species with good dispersal ability, and (iii) overall reduced number of life forms, so that higher diversity of life forms occurs in less altered environments. Individual sizes also differed, so that larger species (body length > 11 mm) were the predominant size category in ruderal biotopes.

Bohac (1990) devised an index of staphylinid communities to evaluate the extent of human interference, based on relative representation of three ecological groups with differing biologies. These groups are as follows.

Table 2.1 The major life-form systems for adult Staphylinidae.

Class	Subclass	Group
Zoophages	Epigeobios	Epigeobionts, walking, large (*Staphylinus*)
		Epigeobionts, walking, small (*Philonthus*)
	Stratobios	On soil surface and in decaying litter (*Othius*)
		In decaying litter (*Medon*)
		In decaying litter and under bark (*Dinaraea*)
		Bothrobionts (*Quedius*)
		Troglobionts (*Domene cavicola*)
	Geobios	Geobionts, running/grubbing (*Phytosus*)
		Geobionts, edaphic (*Geostiba*)
	Psammocolimbets	Coastal (*Stenus*)
		Light and sandy soils (*Astenus*)
	Petrobios	*Domene*
	Torphobios	*Gymnusa*
Phytophages		Dendrochortobionts (*Eusphalerum*)
		Coastal (*Bledius*)
Saprophages		Living in decaying litter and soil (*Omalium*)
		Epigeobionts, small (*Oxytelus*)
		Troglophiles (*Ochthephilus*)
Mycetophages		*Gyrophaena*
Myrmecophiles and Termitophiles		Symphiles (*Atemeles*)
		Synechtres (*Lamprinodes*)
		Synoecentes (*Thiasophila*)

Source: After Bohac (1999) with permission.

- Group R: species remaining as remnants from past natural communities and so representative of those specialized biotopes.
- Group A species: occur in both natural and managed systems (in Bohac's example, forests).
- Group E: eurytopic species that can occupy deforested sites and anthropogenic environments.

The index reflects the extent of human interference within a range from 0 (when only eurytopic species occur) to 100 (only Group R, showing communities unaffected by human activity). This analysis was extended to identify parameters of the critical stages of staphylinid communities, when their structure becomes unstable through change. This arrangement is indicated by Table 2.2, which lists a suite of features that mark departure from natural situations.

In essence, Bohac (1999) believed staphylinid communities to be influenced strongly by the form of landscape change and use, to yield very different characteristics in natural and anthropogenic arenas. Clough *et al.* (2007) also found a functional group approach informative in interpreting differences in staphylinid assemblages between organic and conventional wheat fields, with activity of

Table 2.2 Some parameters indicating the critical stages of staphylinid communities.

Parameter	State
Frequency of ubiquitous specimens	> 90%
Index of community*	< 35
Number of life forms	< 4
Frequency of large individuals (IV, V)†	> 20%
Frequency of individuals with summer activity	> 40%
Non-flying species	Absence
Frequency of species with higher temperature requirements	> 70%
Frequency of species with lower temperature requirements	> 70%
Value of sex ratio index	> 10% from 1 : 1

* The index of staphylinid communities (IS) is defined as

$$IS = 100 - (\textstyle\sum_{i=1}^{n}E + \sum_{i=1}^{n}A)$$

where the first right-hand sum comprises the percentage abundance of individuals of eurytopic species (group E) and the second the abundances of individuals of species of natural and managed forests (group A). Values range from 0, where only eurytopic species are present and the community is heavily affected by human activity, to 100, where only more specialist species are present and the community is unaffected by people.
† The largest of the five size groups of Staphylinidae: group IV are 7.1–11 mm body length; group V are > 11 mm body length.
Source: Bohac (1999) with permission.

predators higher in conventional fields and of detritivores higher in organic fields. Separation of the 174 species of rove beetles in that survey into the main trophic groups (namely predators, parasites, detritivores, fungivores, myrmecophiles, phytophages and unknowns; cf. Table 2.1) enabled detection of patterns not previously found. Raising the potential for using Staphylinidae more extensively as indicators will depend on progress with taxonomy and ability to recognize species unambiguously, and understanding the sampling biases in this very diverse group. The beetles are virtually ubiquitous, and manifest many of the expected general attributes of putative indicators. Many are indeed active, as their common name of rove beetles attests, and pitfall trapping is the most frequently used sampling method. As a step towards determining the adequacy of pitfalls in assessing population density and species richness, Berlese funnel extractions of turf samples from the same sites were examined (Buse & Good 1993). For 20 of 21 sites compared, the turfs yielded species not taken in the pitfalls, but the species pool (119 species across nine subfamilies) in that survey indicates the potential for interpretation of diversity on a forest in north-east England.

One major interpretative problem for assessing indicators continues to be that most authors have slightly (or sometimes radically) different concepts of the term 'indicator' (or may fail to appreciate the different ways in which the term can be applied and various meanings confounded; McGeoch 1998), so that any differences in composition or abundance across treatments may be classed as indicative in some form or another. Criteria may simply be change in richness or abundance, but also turnover (change in species composition), because a common consequence of environmental change is that beetle assemblages lose specialist species but may gain more generalized species not as heavily influenced by the change, or resident generalists increase in abundance. To the non-initiated observer, samples from such assemblages still contain plenty of beetles. Sound interpretations must heed such compositional changes but often they do not, so that analyses only of categories above the species level are likely to mask such tendencies. Interpretative problems increase in faunas with large numbers of undescribed species but because beetles exhibit many of the general features desirable in indicator groups (diverse, widespread, easily sampled, functionally important and ecologically varied, with short generation times leading to rapid responses to changes, and proven responsiveness to many kinds of imposed change), they continue to be advocated, even though the reasons for observed change may be unknown. As Andersen (1999) emphasized, such understanding is necessary to help overcome widespread but uncritical advocacy for use of many insect groups as preferred indicators. However, whilst some claims of beetle values as indicators may be exaggerated by entomologists seeking to promote the values of their favourite group, a comment on the values of Carabidae by Rainio and Niemela (2003) – 'carabids are useful bioindicators, but as crucial understanding of their relationships with other species is incomplete, they should be used with caution' – applies at least equally to most other beetle families. A perceptive evaluation of values of Carabidae as biodiversity indicators (of effects of forest management) by Work *et al.* (2008) is a salutary caution against over-generalization. Ten separate studies in Canadian forests (four in the western half of the country, the other six eastern), all undertaken to measure changes in carabid assemblages in response to disturbances, were evaluated using a pool of 152 species (of which 16 species, 12 of them transcontinental, comprised 90% of relative abundance, and 45 were represented at one site only) across a total of 365 sampling sites; this represents perhaps the largest-scale comparison yet made on this theme. Differences in assemblage composition were evaluated in relation to location (province), silvicultural treatments/variables and dominant cover type, with use of indicator-species analysis (Dufrene & Legendre 1997) to evaluate individual species' responses. The major finding was that carabid assemblages indeed responded consistently to disturbances, but the responses of individual species and detailed changes in species composition reflected the context of regional geography and individual fine-scale differences among forest ecosystems. In this comparative survey, carabids appeared to be suited better to fine-scale (stand-level) disturbance indications, but not as consistently suitable for large-scale regional or national monitoring. Regional differences in ecosystems and carabid assemblage composition then limit the values of individual species as large-scale indicators because the comparative influences of these different wider contexts are scarcely understood. Work *et al.* (2008) also emphasized the need for programmes planning to monitor such responses to be sufficiently extensive to be able to detect the responses of rarer species, rather than simply relying on the more abundant, probably ubiquitous and relatively generalist species, whose responses, should any occur, may be far less informative of conservation need. The large area of Canada is likely to reveal such variety clearly, whereas greater uniformity might occur in smaller areas, such as some parts of Europe.

However, the heterogeneity of management practices in forests or of other vegetation implies that considerable care is needed in transferring implication of indication from one site or stand to others. As Spence *et al.* (2008) emphasized, suites of indicator species (beetles or others) must be regionally specific, and there is no real substitute for species-level assessments of responses. Single species' responses may differ from those of even their close relatives, or from the assemblage as a whole so that, for example, saproxylic beetles as a group may not be associated directly with fungi sometimes used as a continuity indicator species of old-growth forests important for conservation (Norway; Sverdrup-Thygeson 2001). In general, these showed no correlation with beetle species richness or with the presence of red-listed beetles in the forests. However, most of the last group were found in only very small numbers, so the interpretation may be open to revision. The cryptophagid *Atomaria alpina* was correlated positively with numbers of indicator logs on a site, whereas the numbers of red-listed Cisidae showed negative correlation with those logs. Rainio and Niemela (2003) discussed cases for the value of carabids as indicators of disturbance and as surrogates for wider biodiversity as biodiversity indicators, a role in which (as suggested above) they at times seem to fall short. The great biological variety of Carabidae renders generalizations over their roles as disturbance indicators difficult. Many carabid assemblages include high proportions of generalist species to comprise a stable and unresponsive core suite, but also more responsive specialist species that may individually increase or decrease with changes in vegetation, soil moisture, fragmentation of habitat, and management regimes imposed to forest or grasslands.

Likewise, strong cases have been made for using carabids as indicators in cropping systems, but Holland *et al.* (2002) questioned whether they are indeed effective surrogates, regarding it as unproven whether they can be used to predict influences of farming systems on either a wider array of arthropod taxa or the total community of ground-active arthropods. They emphasized, as is the case also in many other such claims and extrapolations, that the ecology of the other macroarthropods has generally been much less thoroughly studied, so that the causes of their responses and changes can generally only be inferred. As Hance (2002) also noted, many studies involving carabids as putative indicators have been inconclusive. Possible sources of confusion, all of which would apply to other groups, have included (i) scales of treatment effects, such as working in small plots; (ii) sampling commonly by using pitfall traps alone; and (iii) insufficient taxonomic penetration masking influences on individual species. In addition, simple analyses might not reveal the relatively subtle responses detectable by multivariate analyses of the species composition, and responses by dominant species may mask those of others. The rationale for investigation presented by Holland *et al.* (2002) takes a different perspective. Rather than implicitly accepting that pitfall trap catches can detect long-term experimental treatment effects and then analysing the catches to determine whether such effects have occurred, they opted simply to accept that, whilst limitations to this technique occur, its convenience and simplicity ensure that it will continue as a mainstay sampling method. The need is then to see whether it can detect long-term treatment effects, or in their study effects of specific pesticide regimes, and to compare responses of Carabidae to those of other farmland arthropod groups. Their major trial, undertaken over 6 years with pitfall traps, yielded 241,666 arthropods, dominated by Carabidae (50%), Staphylinidae (18%) and Araneae (32%) – this predominance alone providing very solid grounds for selecting these groups for investigation – across two farming systems and incorporating six sites and a total of 96 plots, each of more than 5 ha.

The results for Carabidae revealed considerable inherent variation between sites but also between fields so that substantial numbers of replicates may be needed to validate observed differences. Concluding from this substantial survey, and also from trials on pesticide effects, Holland *et al.* (2002) suggested the following practical outcomes as leads for future work and greater understanding.

1 Pitfall trap catches of Carabidae are not likely to detect long-term effects of farm management practices, other than when high use of organophosphate insecticides occurs.
2 Response of the carabid community to farm management practices with high use of organophosphates could be suitable as an indicator of the greater macroarthropod community response, with even short-term changes having implications for prey availability to consumers.
3 Monitoring carabids rather than all macroarthropods would benefit overall interpretation if the effort saved could be diverted to also monitoring some microarthropods (Collembola), which are more sensitive detectors of some pesticide effects.
4 Excluding the dominant species of Carabidae from analysis affects the outcomes of the analysis. It is relevant to undertake community analyses with and without those species included, to determine their overall effects.
5 High heterogeneity of carabid diversity and abundance, sometimes over small scales as well as between localities, may mask subtle responses to local conditions. Individual species or diversity indices have only limited value for indicating impacts of farm management regimes, so that wider community-level analyses may be more informative.

A starting point for evaluation in any previously unstudied context may be simply that particular beetles are broadly recognized as important components of a system, as Michaels (2007) noted for Staphylinidae and Tenebrionidae in Australia, with this importance enhanced should indicator values have been demonstrated elsewhere in the world. For these two families, evidence of indicator value beyond the responses of individual species to particular disturbances remains ambivalent.

A practical appraisal of the value of a suite of South African dung beetles as indicators (McGeoch *et al.* 2002) used the IndVal (Indicator Value; after Dufrene & Legendre 1997) approach. This combines measurement of the extent of habitat specificity and its fidelity with that habitat in order to explore and clarify possible indicator values. The index is calculated independently for each species, and the habitats (or sites) can be categorized or grouped in any consistent way to allow for different scales of appraisal. Species with high specificity and high fidelity (shown by frequency of occurrence across sample sites) will have higher indicator value (Fig. 2.1) and are often also abundant.

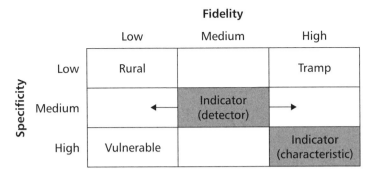

Fig. 2.1 General analysis of indicator parameters: species characterized by combination of extent of environmental specificity and fidelity, and classified as either characteristic indicators, detectors, tramp, rural or vulnerable species. (After McGeoch *et al.* 2002 with permission.)

As McGeoch *et al.* (2002) noted, abundance (and, in consequence, fidelity) may vary over time in two important ways, (i) with season and weather and (ii) through disturbance-induced changes, which are likely to affect the species present differentially. The most sensitive species are commonly among the most relevant indicators but, because of their vulnerability, may also be those of greatest conservation concern. However, McGeoch and her colleagues also discussed the practical differences between two major categories of indicators, as shown in Fig. 2.1. At the bottom right of the figure, high-specificity/high-fidelity indicators are restricted to a single ecological state so, notwithstanding changes in abundance, may not reveal much of the form of ecological change. In contrast, 'detectors' as medium-level species may be more informative because they transcend ecological states with varying levels of fidelity, and facilitate long-term assessments. Following McGeoch (1998), this scenario was explored from assemblages of dung beetles in Tendo Elephant Park, South Africa.

Occasionally, particular beetles have become embroiled as threats to other organisms of conservation concern. Larvae of the native North American elaterid *Landelater sallei* damage eggs of the loggerhead turtle (*Caretta caretta*) in Florida, and this species was the most damaging predator of eggs found by Donlan *et al.* (2004), who suggested that the beetle might be a threat also to other marine turtles nesting in the USA. On the Galapagos Islands a scarab (*Trox suberosus*, given as *Omorgus suberosus* by Peck 2006) feeds on eggs of turtles (found in 96.7% of turtle nests on the Galapagos, and causing up to 21.6% egg mortality; Peck 2006) and the Galapagos land iguana (*Conolophus subcristatus*) (Allgower 1980), and there are several records of Tenebrionidae and Elateridae eating turtle eggs in the eastern Mediterranean.

Planning for species conservation

This individual-species level of conservation is the first way in which many people view practical conservation, and has been the major driver of moves to allocate species as threatened in some hierarchical way to aid setting priorities for action. Any tangible threat provides a basis for amelioration and conservation management. As noted earlier, this endeavour commonly arises through some listing and this step may even be a formal prerequisite in gaining access to funding or other agency support. Almost universally, and despite the fact that listing draws attention to the plight of many very worthy beetle species, such listings beyond parts of Europe are very incomplete, so that published directories of taxa of conservation concern must simply be regarded as indicative of a far wider array of equally deserving beetles. Subsequent selection from lists of the fewer species to be given priority is also difficult, with numerous criteria proposed to help rationalize and advance the process, within the general acceptance that choice or triage is indeed needed.

The units

The entities involved are, in the main, taxonomic species but with no precise definition of 'species' included in most regulatory or advisory documents. Some legislations also recognize subspecies as equivalent to species, if these are recognized by expert consensus, and a few also provide for recognition of 'significant populations'. This category includes evolutionarily significant units, to cater for genetically or geographically unusual units that are not named formally. Entities below full species can easily become vague; as Pearson *et al.* (2006, p. 17) put it for North American tiger beetles 'For conservation efforts, subspecies names formalize distinct variations within a species that can make policy decisions more palatable to politicians and legislators' but also 'The key to using subspecies . . . is never to forget how imprecise they are'.

Many taxonomists have their own ideas about definitions of subspecies, but the common elements to most of these are (i) consistency of recognition, supported by distinctive characteristics that give the entity a unique definition that separates it from all related entities; (ii) a unique habitat or geographical range, usually segregated from that of related entities; and (iii) biological features that also segregate it as a unique entity. In most cases, the genetic data needed to confirm evolutionary distinctiveness and relationships are not yet available.

In addition to these formally recognized and named categories, many insects exhibit considerable within-species (or within-subspecies) variation so that, for example, colour patterns of individual Cicindelinae or Coccinellidae can differ substantially within a single population and the frequencies of those patterns (which in many species are sufficiently discrete and stable to be termed 'morphs') differ across the species' range or in different populations. Such varieties may reflect environmental influences, genetic heterogeneity or unknown causes. Characterizations of these do not usually provide for formal legal recognition but, of course, may contribute significance to isolated entities. However, variations can augment the interest of particular beetle groups to collectors and occasionally such variable populations or those with desirable morphs may be threatened by over-exploitation. In a few species, colour variations are linked with pollution effects, leading to value as indicators: the industrial melanism of the ladybird *Adalia bipunctata* is perhaps the most famous example of this.

The recognition of evolutionarily significant units or significant populations is receiving increasing attention in setting conservation priorities at this fine-grain level, backed by recognition that conservation of genetic diversity is an important aspect of conservation need. It links strongly with the tracing of evolutionary transitions and distribution patterns termed 'phylogeography', and the following beetle examples simply demonstrate some aspects of conservation application.

1 *Prodontria* in Central Otago, New Zealand. The incidence of two colour forms of *Prodontria* (Scarabaeidae) near Alexandra, not far from the Cromwell site of *P. lewisii* (see p. 57), led to debate over whether two species were represented. Genetic studies examined variation within and among populations (Emerson & Wallis 1994), and implied the existence of a single encompassing polytypic species, but much of the diversity revealed was between or within populations rather than between the colour morphs. This suggested that the systematic value of those morphs had little formal significance. The two forms are now recognized together as *P. modesta*, which occurs only over about a 10-km radius from Alexandra, with a small area of morph sympatry (Barratt 2007). The species is of interest in demonstrating the possible confounding interpretations of visual morphs and genetic diversity, with the beetle's grassland habitat threatened by horticulture, viticulture and development and increasing its immediate conservation interest.

2 Some coastal Cicindelidae in Japan. With increasing habitat fragmentation, much beetle conservation is necessarily concentrated on isolated populations. Genetic appraisal to assess the distinctiveness of such populations helps to illustrate their evolutionary significance and the patterns of their evolution and isolation. However, generalization is often difficult. For example, four Japanese tiger beetle species examined by Satoh *et al.* (2004) had different evolutionary histories since they originated in continental Asia and reached Japan across land bridges during the Pleistocene. All four species are red-listed in Japan, and one (*Abroscelis anchoralis*) is regarded as critically endangered. Satoh *et al.*'s mitochondrial DNA studies revealed that the three distinctive regional populations of *A. anchoralis* have persisted without contact for about 0.2 million years. One was clearly considered an evolutionarily significant unit: it is restricted to a single beach, about 2 km long, in Ishikawa Province and no nearby populations are known. In contrast, populations in south-east Kyushu occur along a stretch of coast almost 135 km in length and are much less vulnerable. The third unit (Tangea-shima Island) needs further investigation but, at present, conservation attention to this species can clearly focus for priority on only one of the three areas where *A. anchoralis* occurs.

3 *Pimelea* on the Canary Islands. Genetic studies of the species of *Pimelea* (Tenebrionidae) on the Canary Islands (Contreras-Diaz *et al.* 2003) exemplify how such work may contradict, complement and clarify conventional morphologically based taxonomy to lead to clearer recognition of significant units and aid in conservation focus and priority. One of the species, *P. fernandezlopezi*, is of particular conservation interest because it is known from

a single population on a fossil dune on the island of La Gomera and is threatened by habitat changes. Several other species have also been cited on the Canaries list of endangered species and/or the Spanish list of endangered species. La Gomera is separated from Gran Canaria by about 130 km, with deep water between them, and has an independent volcanic origin. *Pimelea fernandezlopezi* is likely to have arisen from an early colonization from Gran Canaria. The three endangered species on the larger island are all endemic, and threatened by habitat changes from expansion of tourist resorts, sand extraction, dumping sites and related developments, so that many of their populations are now genetically and geographically isolated through fragmentation. Current conservation measures do not include genetic considerations (Contreras-Diaz *et al.* 2003) and such expansion of scope, as would be the case also for many other beetle conservation efforts, might help tighter focus for deployment of limited resources.

Evolutionarily significant units (ESUs) reflect separation of populations within a species, with the extent of those differences determining perceived conservation significance. Vogler and DeSalle (1994) recommended that ESU should be used as a technical term for clusters of organisms that are evolutionarily distinct and hence merit separate protection, and this sentiment underpins much modern usage, incorporating consideration of different haplotypes and their association with morphological and ecological differences. Thus the North American Puritan tiger beetle (*Cicindela puritana*) (see p. 141) occurs in two distributional centres, where the beetles are diagnosably distinct genetically and regarded as separate ESUs (Vogler & DeSalle 1994). The habitats also differ considerably: the Chesapeake Bay populations occupy coastal cliffs (20–30 m high), whereas those on the Connecticut River occupy sandy patches in river bends, so that clear ecological differences also occur.

In some other cases of need to distinguish between putative subspecies, genetic work has been signalled as necessary to help resolve taxonomic ambiguities, but has not yet commenced. One such case is the two subspecies of *Desmocerus californicus*, riparian cerambycids in California (see p. 192), with the need to unambiguously define the threatened Valley elderberry longhorn beetle (see p. 196).

The process

Any entity selected for conservation attention, by whatever means and of any taxonomic or more informal status, must become the focus of a practical programme to investigate its needs and manage it (as far as possible) to assure its well-being. Selection is thereby a responsible act, often with subsequent formal commitment that draws on very limited pools of finance and expertise. Management may target the species across its entire range, or address the populations in only part of that range, as in particular countries of Europe or individual states of the USA or Australia. Each such political entity, with the principle extended to smaller administrative units or jurisdictions, may have its own list of priority species, and this may reflect that a beetle is threatened in some parts of a range but not in others. A universally threatened beetle may be distributed narrowly or more widely, in which case the range may transcend political boundaries and lead to possibility of wide cooperation for its conservation. Narrow distributions, commonly explained as narrow-range endemism, are often difficult to evaluate or interpret. Conservationists may find it tempting to consider them as remnants of formerly wider ranges, resulting from anthropogenic causes and now indicating conservation need. However, some such distributions are indeed natural. The New Zealand Cromwell chafer (*Prodontria lewisii*) has been sought extensively in Central Otago since it was described in 1903. However, Watt (1979) suggested that it may never have occurred naturally beyond the 500-ha area of the Cromwell sandy loam dune system, within which the beetle now occurs only in a single 81-ha reserve (Barratt 2007). This reserve was formalized through the efforts of Watt in the 1970s, when basic needs included fencing to exclude unauthorized vehicles and control of woody vegetation, in this case predominantly of self-seeding *Pinus radiata* from nearby plantations. The surrounding countryside is now apparently unsuitable for the beetle, and threats to *P. lewisii* include changes in vegetation and predation by vertebrates (such as owls), so that recent surveys (Barratt 2007) have confirmed the very restricted distribution and small population size of the chafer. Parallel studies are needed to elucidate the status of other localized species of *Prodontria* in New Zealand.

Conservation management for the Cromwell chafer exemplifies a common situation for narrowly endemic insects, namely that conservation must focus on particular sites, often small, as the only place(s) where the species is known to occur. Although it may be possible to found other colonies (see p. 140), the original sites are likely to remain of paramount concern. Management must then inevitably become site-specific and the species may well become conservation-dependent because, without continuing management, the site may change and diminish in suitability. Such programmes easily become expensive, and necessitate long-term commitment.

At any geographical scale, conservation must be well planned and management must have clear objectives (New 2009) with the means to pursue those defined. The formal act of listing a species as threatened or for conservation priority may oblige further investigation and/or the development of some formal document or plan. Further research will almost always be needed to determine conservation status and needs fully, with such work targeting major gaps in knowledge on which to found management. In some instances, this work may reveal that the beetle is more secure (such as by being more widely distributed, with larger populations, and less threatened) than supposed at the time of listing: this is a very positive outcome of selecting the species for attention, and under some legislation it can then be delisted quickly to allow more attention to others.

Any selection of a species for individual conservation presupposes sufficient biological and distributional knowledge to render that selection reliable and convincing. The common lack of that knowledge, particularly in faunas with a poor taxonomic framework, may lead to wider approaches as a prelude to such individual focusing. In New Zealand, McGuinness (2002, 2007) set out the major considerations in *Threatened Carabid Beetles Recovery Plan*, a flagship document for furthering insect conservation in the country. Slightly more than 200 beetle species, including 48 Carabidae, are listed as threatened or possibly threatened in New Zealand, with predation by introduced mammals the most commonly cited cause of concern. The recovery plan covers 56 taxa, and these are divided into four groups.

- Group A comprises beetles with recovery plans and requiring active management. The four species listed as needing specific recovery actions are *Zecillenus tillyardi*, *Mecodema costellum costellum*, *M. laeviceps* and *Megadromus* Eastern Sounds, listed under the approach that although reasons for decline are still not understood fully, actions must precede conclusive evidence in order to reduce chances of imminent extinction – in other words, the precautionary principle is an effective guide for instituting conservation. These four species were sufficiently well understood to provide an educated guess as to causes of decline and the management steps needed. Thus, for *M. c. costellum* on Stephens Island, lack of fallen logs as refuges, resulting from forest clearance, might limit the beetles' ability to increase in numbers, so that placement of old wooden fenceposts was undertaken. These posts have been frequented by the beetle, and population monitoring is continuing. All four of these priority species have received some level of conservation attention.
- Group B comprises beetles requiring survey or other information gathering. The 24 species listed here were supported by sufficient information to indicate declines, but further survey and monitoring were needed to establish and confirm the extent of loss. The species were categorized as high, medium or low priority.
- Group C comprises beetles requiring taxonomic revision or better means for identification: 11 species needed additional work to establish their taxonomic validity before effective conservation measures could be contemplated.
- Group D comprises beetles listed previously as of conservation concern or need but which have now been downgraded and are not proposed for any current conservation action.

Part of the role of this plan was to help overcome the gaps in knowledge of individual species by placing each in wider context through which priorities might be sought. McGuinness (2007) set out a series of key questions to help focus aims further, and drawing on the approach for carabids.

1 What species do you want to protect (prioritization)?
2 What is causing their decline?
3 Can we manage the agents of decline?
4 What are the key sites (prioritization, including considering areas which will conserve the greatest number of threatened species)?
5 Does it constitute a priority for funding (prioritization)?

This sequence implicitly requires that priority on threat status alone does not always align with species that can be managed, and reveals the almost universal need to undertake further research on beetles as a basis for specific conservation management. The insights from McGuinness' schemes led to a system for decision-making of very wide relevance in insect conservation (Fig. 2.2), and help to emphasize that any such management plan must be a responsible and realistic basis for management. However, one practical limitation of the New Zealand system, widespread also elsewhere, is that it is a voluntary guide available for people to use if they wish to do so, but is not enforceable and

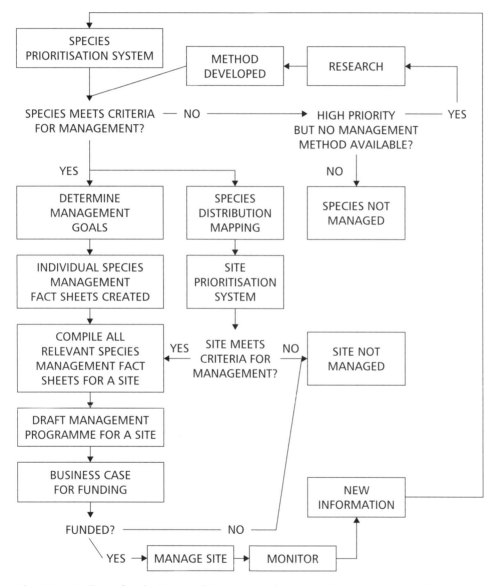

Fig. 2.2 Outline of a decision-making process for insect site management for conservation. (After McGuinness 2007 with permission.)

carries no accountability. Nevertheless, the New Zealand carabid plan is the only national one for New Zealand covering any group of beetles in this way, and demonstrates the Department of Conservation's intention for practical conservation not to be delayed unduly once a need is perceived.

The comprehensiveness and coverage of beetle species management plans vary considerably in different places, and with the particular purpose or focus in relation to both driving regulation/legislation as the impetus for production, and the nature of the receiving readership. Many UK plans for priority species, for example, are short statements of need couched in rather general terms but sufficient to encapsulate concerns and priorities within the context of a well-known fauna, and with the main readership knowledgeable and able to provide considerable informed and sympathetic support for conservation. Thus, for the charismatic stag beetle *Lucanus cervus* (see p. 39), the national plan could refer to numerous more local plans each contributing to the wider conservation endeavour and appreciation. In contrast, some US plans constructed following species listing under the Endangered Species Act are much more comprehensive, and may extend for several hundred pages to incorporate summaries of virtually all available biological and distributional information, with minutiae of proposed remedial or management measures for each individual site involved. For many such species, the primary readership (such as conservation managers who are expected to pursue those plans in the field) will not be entomologists, and such detailed background is invaluable in setting perspective. However, any conservation plan should ideally encapsulate a variety of practical topics (New 2009) to indicate uncertainties in biological understanding and interpretation of conservation status and need, the major research and management objectives, the actions needed over particular time scales and ways in which the success (or otherwise) of the endeavour can be monitored and evaluated. New (2009) emphasized the accountability and practical values of having SMART objectives in insect conservation plans, so that firm guidelines are agreed at the outset, with the actions designed to achieve these objectives being Specific (unambiguous), Measurable (criteria and duration specified), Appropriate (related to the overall goal of the plan as stated in the initial broad objective or mission statement), Realistic (achievable within the time and resource parameters specified), and Time-based (with a cut-off time for completion). Collectively, these are difficult to design but, without doing so, there is danger that any management plan will become simply an open-ended 'wish list' without assessment of suitability or accomplishment. In essence, any management plan, for a beetle or other species, must be persuasive to three major readerships, many of whom will not have any detailed knowledge of the species and who may find the very idea of insect conservation novel or even risible. These groups comprise those who have an administrative or legal responsibility for conservation; those who will be funding and implementing the plan; and those who want to know what is being done and why. As Burbidge (1996) noted, the general requirements of such plans should (i) enable relevant conservation work to be based on and guided by the best possible information, as well as by focused objectives and actions; (ii) maximize the possibility of success and minimize the chances of the species becoming extinct; and (iii) inform the public over what is being done to save the species.

For most beetles there will be considerable initial uncertainty over what to do, and an interim set of tasks may be needed in order to render the wider plan feasible. Not least, many species-focused exercises are likely to occur on particular sites, and the future security and management restrictions for those areas is thereby of critical importance. Despite the recognition of such fundamental needs, many conservation plans are not couched in such formal terms as suggested above but, equally, are open to suggestion that they may consequently be less practicable. In some cases, lack of detail is inevitable, because the biological or distributional knowledge needed to formulate precise actions is simply not available, so that the document becomes a statement of intent and goodwill founded on the best information to hand. Again, parallels occur in (and lessons can be learned from) plans designed to counter pest beetles, for which biological understanding is often facilitated through substantial economic impacts and importance, so that control becomes an urgent and politically accepted issue, supported and driven far more determinedly than many conservation projects. The mountain pine beetle *Dendroctonus ponderosae* has caused massive recent losses to commercial timber supplies (lodgepole pine *Pinus contorta*) in British Columbia, with the most recent epidemic reported as killing around 283 million cubic metres of timber (Government of British Columbia 2005). An action plan included seven objectives, reflecting coordination between several relevant government ministries as well as industry and other stakeholders, and recognizes the massive economic and social impacts on human communities dependent on the timber industry. The overall goal is 'to sustain long-term community, economic and environmental well being while dealing with the short-term consequences of the epidemic', and the objectives were adopted to guide activities towards this. These, noted in Table 2.3, were accompanied by proposed more specific activities, with biological needs including how to predict beetle build-up and how to develop soundly based mitigation responses.

For most non-pest beetles, equivalent levels of practical support are much more difficult to realize and it is thus important that the efforts made are indeed targeted as effectively as possible. Much of that effort devolves on clarifying

Table 2.3 Countering the epidemic of mountain pine beetle (*Dendroctonus ponderosae*) in western Canada: the major objectives of the Action Plan (2005–2010) to mitigate its effects.

Encourage long-term economic sustainability for communities affected by the epidemic
Maintain and protect public health, safety and infrastructure
Recover the greatest value from dead timber before it burns or decays, while respecting other forest values
Conserve the long-term forest values identified in land-use plans
Prevent or reduce damage to forests in areas that are susceptible but not yet experiencing epidemic infestations
Restore the forest resources in areas affected by the epidemic
Maintain a project management structure that ensures coordinated and effective planning and implementation of mitigation measures

---

Source: Government of British Columbia (2005).

Table 2.4 Objectives of the recovery plan for Hungerford's crawling water beetle (*Brychius hungerfordi*).

Determine and ensure adequate population size, numbers, and distribution for
   achievement and persistence of viable populations and long-term survival
Identify habitat essential for all life stages and ensure adequate habitat conservation
Identify whether additional threats exist

These objectives will rely heavily on researching the species' biology and habitat
requirements so that we may more adequately assess and alleviate threats and
develop measurable and objective Recovery Criteria

Source: Tansy (2006).

and interpreting biological information, and specifying the major gaps that should
be addressed. Thus, little was known about the precise ecological needs of
Hungerford's crawling water beetle at the time the major recovery plan for this
species was written (Tansy 2006), so that a preliminary research programme
was needed to 'gather more information on the species' life history, habitat
requirements, distribution and ecology in order to determine if this species has
inherent biological constraints'. However, those research needs were set out clearly
in the recovery plan (Table 2.4), mainly in very general terms and without time
lines. Nevertheless, the recovery plan also contained clear interim recommendations for action, in conjunction with the research needs, and stated recovery
criteria.

   More generally, it is very easy for such stated research requirements within
a management plan to become open-ended and lose strict relevance to practical conservation needs: research on a threatened beetle is not automatically
conservation biology or management and, whilst any additional knowledge may
have some relevance, any research should preferably test hypotheses on specific
management themes or objectives, or address key biological questions for
which answers are needed to further conservation. An unusual (and valuable)
feature of Tansy's (2006) water beetle plan, and some other plans prepared by
the United States Fish and Wildlife Service (USFWS), is inclusion of comments
made by reviewers of the draft, and the author's responses to these, in part of
which it is stated that 'a key component of the recovery plan is implementation
of a research programme, which will help us evaluate the reliability of our
assumptions', with these assumptions set out clearly within the existing biological framework.

   General themes that may need specific investigation or clarification for
planning to manage any beetle species of conservation interest include the
following.

1   Clarify conservation status against the pertinent or locally dictated criteria,
    with this status treated as dynamic but incorporating aspects of distribution,
    abundance, population size and number, evidence of decline and the causes
    of decline as a basis for management; linked with this, clarify the life history
    and pattern of seasonal development of the species.

2   Make a clear decision about whether conservation measures should aim to retain the status quo (i.e. to conserve existing populations at their current levels and range) or to increase the size, number and distribution of populations, i.e. more aggressive recovery.
3   Determine the optimal conditions for the species' well-being, in terms of critical habitat (see p. 77) and resources, and how those conditions may be supported or managed and resource supply enhanced and assured; determine population structure of the species and the role of dispersal in maintaining population structure and dispersion across the landscape.
4   Eliminate or mitigate all known current and anticipated threats.
5   Determine whether site or habitat security can be assured as a basis for long-term management; if not, determine whether alternative approaches, perhaps involving translocations (see p. 138) or *ex situ* conservation measures, are needed and feasible.
6   In conjunction with all of the above, assure provision for effective and continuing coordination of the plan, including periodic review based on adequate monitoring, and for management to be dynamic and responsive; for all data to be archived responsibly; and for all resources needed (including finance and personnel) to be sought and assured, with clear allocation of responsibility for each specified duty.

The agenda for any management plan will need to be planned very carefully, together with setting priorities and determining the sequence and duration of all actions proposed. Simply discovering what is known (and, by complementarity, what needs to be known) will usually be among the initial tasks, accompanying investigation of site tenure and security. At this stage, commonly following listing of the species as the formal impetus for conservation, independent peer review of the information is important. The two major outcomes of a listing and preparation of a management plan for a beetle are (i) that further appraisal and/or field study reveals the species to be more secure than previously believed and not in need of particular conservation management, or (ii) that management for sustainability or recovery is indeed needed and can be undertaken through a defined programme presented in the plan. Should the first option occur, no further action may be needed, other than possibly some watching brief, and the species can be downgraded rapidly in importance to allow resources to be deployed to more deserving taxa.

Population structure and beetle dispersal

A small pond, an isolated dead tree, an individual bracket fungus on that tree, an ant nest or a small patch of cleared ground in a forest are all to some extent isolated islands in the landscape, and (as for numerous parallel examples) must be sought by beetles that depend on them. Many are maintained naturally – trees are continually dying and fungi growing on them – so that the resource becomes dependable and, in evolutionary terms, predictable. The lifestyle and dispersal capability of organisms can be honed towards finding and exploiting such patchy or widely dispersed breeding environments, but their availability and accessibility may be reduced by human activities, a core cause of conservation need. Much conservation attention to beetles has focused on assuring the continuity of populations on particular sites, but with little knowledge of whether the presumed habitat or site fidelity or long-term residence is really a natural condition. Thus, some large ground beetles in Europe, contradicting the implications of their common group name, climb trees, so may occasionally be found unexpectedly in surveys of arboreal insects. Weber and Heimbach (2001) discussed possible reasons for this in *Carabus auronitens*, and suggested four possible reasons for tree-climbing, namely avoiding predation on the ground, seeking optimal microclimate at particular times, seeking food as predators, and as a consequence of interspecific or intraspecific competition.

Interpretation of insect population structure was long confined to the contrast between so-called open populations, dispersing freely across the landscape and with local changes in abundance including the influences of immigration and emigration, in addition to natality and mortality; and closed populations, those isolated from others by distance, physical barriers or specific resource distributions and within which changes in numbers reflect internal processes only. The latter has been the classic focus of conservation, because changes in distribution and abundance over time may be related to specific local conditions and change, and declines raise concerns for sustainability and survival. However, the convenient simplicity of this dichotomy is largely unrealistic, and the implications of metapopulation structures and dynamics change the picture considerably. With this approach (Hanski & Gilpin 1991; Harrison 1994; Hanski 1999 for historical and conceptual background), small and apparently isolated populations (as metapopulation units) within the landscape may be independent demographically for much of their existence but subject to occasional immigration or emigration. Each may also become extinct as part of normal existence, so that the metapopulation range of the species includes suites of largely independent habitat patches, each subject to series of rolling extinction–colonization cycles reflecting their condition, but implying that an individual local extinction may be entirely usual as part of normal population dynamics rather than always cause for crisis management. Den Boer (1987) suggested that, with continually changing conditions, the mean survival times for local populations of Carabidae may be naturally only a few decades, and rarely more than a century.

Dispersal then becomes a critical process to appraise in assessing the level of population isolation or whether changes at landscape level may influence the likelihood of maintaining a metapopulation structure. It reflects the relative benefits of staying or moving, with one premise underlying ideas of metapopulation structure being that local extinctions are followed by colonizations and may indeed promote dispersal because the possible benefits of dispersal may be increased if suitable vacant habitat patches/sites are available. It has long been acknowledged that the presence or extent of migratory dispersal in insect life histories is linked commonly with predictability of habitat suitability (Southwood 1962), so that more ephemeral habitats necessitate more frequent dispersal. Resources used by dung beetles, carrion beetles, water beetles frequenting temporary pools, and many parallel scenarios, reflect that only one (or, at most, a few) generation(s) may be passed on that individual resource, and dispersal is needed to track resources for future generations in the landscape. Water beetles are commonly attracted to shiny surfaces, presumably mistaking these for water, for example Welch (1990) reported several species landing on car roofs in England. Indeed, glass panels placed on the ground and inclined at one end towards a collecting container have been used to evaluate dispersal of Dytiscidae, based on their attraction to ultraviolet light (Lundkvist *et al.* 2002). Understanding dispersal, and its relationships with population structure and isolation in relation to dispersion of suitable habitats, is a key aspect of rendering species conservation management effective, in leading to promotion of connectivity (see p. 97) in the landscape. Management may seek to reduce extinction rates for populations, but must also ensure that habitat patches are sufficiently close that recolonization may be possible, should a metapopulation be involved. Rates and distances of beetle dispersal may become important, and species-specific, considerations in conservation. Both vary considerably, and categorizing the movement pattern and ability of any beetle species without reference to the features and context of the dispersal arena (i.e. the landscape) is unwise and a powerful source of error.

One important ecological focus for the study of population structure has been saproxylic beetles, which typically exhibit high dispersal because they depend on a patchy resource that is, at least to some extent, unpredictable in the landscape. However, those beetles depending on dead wood in tree hollows may be occupying a habitat patch suitable for tens of generations, perhaps even centuries, so they may be more sedentary than commonly supposed (McLean & Speight 1993) and depend on relatively infrequent inter-patch movements for genetic exchanges between populations. Gaining information on their dispersal has been undertaken most extensively by trapping beetles arriving at stumps or cut logs, but linking this with life-history information and development may necessitate the twin approaches of using window traps (or others) to sample beetles attracted to the wood (as immigrants and potential colonizers) and emergence traps to sample those leaving it as newly emeged adults (potential emigrants).

The endangered *Osmoderma eremita* (Scarabaeidae) has been studied intensively in Sweden (Ranius 2000; Ranius & Hedin 2000) in order to determine how dispersal may affect the presumed metapopulation dynamics and whether conspecific attraction occurs (see p. 25), in which case aggregations may lower the proportion of occupied host patches within the arena of activity. Beetles were captured in pitfall traps set in the wood mould of tree hollows of ancient oaks, and individually numbered by marking the elytra. Trials over 5 years (1995–99) involved a total of 839 individuals, but only 9 of 901 recaptures were in a separate tree from the initial capture, the greatest distance tracked was 190 m (over a 10-day period) and all inter-tree movements were within the same stands of trees. It seems that *O. eremita* (see p. 84) undergoes more short-range than long-range flights, although there is some (probably unfounded) doubt over whether lack of long-distance flight records reflects beetles having actually moved out of the study area so not being retrievable during recapture sampling. The major practical inference is that the population (or metapopulation unit) associated with each individual tree might become extinct and not be replaced easily, even though the habitat remains suitable. Over the last two centuries fragmentation of forest habitat (see p. 94) has probably prevented dispersal of *Osmoderma* into smaller younger stands, so that long-term population viability may now depend on increasing the size and connectivity of stands with hollow trees (Ranius 2000). Earlier occupancy of individual trees is reflected by the presence of body fragments in the hollows. By studying those fragments, the probability of occurrence per tree increased with the number of hollow trees in a stand, and the frequency of beetles increased with large amounts of wood mould in hollows. A single hollow tree might contain many adults though most trees do not, a possible implication being that *Osmoderma* manifests features of a 'habitat-tracking metapopulation' (*sensu* Harrison & Taylor 1997), with colonization/extinction reflecting successional changes influencing the suitability of the habitat patch (Ranius 2000). *Osmoderma eremita* occurs only rarely in single trees and very small stands, and apparently now exists only in very small metapopulations with few interacting units. The fragmentation that has led to this scenario might also lead to increased extinction rates due to inbreeding (Ranius 2000). Many small populations appear to be at risk, because they rely on isolated patches (of one to a few trees) in the landscape and lack connectivity with other groups. Even when several hollow-bearing trees occur together, the carrying capacity for *Osmoderma* may be limited by only small amounts of wood mould being present (Ranius 2007).

Populations have sometimes been described as patchy when effective connectivity has broken down. According to Harrison (1991) this state occurs when (i) movements between patches are sufficiently infrequent that individuals cannot occupy many patches during their lifetime, but (ii) on average an individual inhabits more than one patch in its lifetime and (iii) local extinctions either do not occur or are not significant because of high dispersal so that suitable patches within the system are occupied. This situation was implied in the riparian carabid *Bembidion atrocaeruleum* (Bates *et al.* 2006), which disperses between habitat patches but mainly over relatively small distances as dictated by the habitat structure.

Mark–release–recapture studies on individual beetle movements and population features have fostered numerous innovations in methodology for marking and tracking, and are the foundation for much of the information on which conservation concerns are founded. Laser marking of carabids can replace paint for both mass-marking and individual code-marking of taxa such as *Pterostichus cupreus* in Europe (Griffiths *et al.* 2005). The initial physical marking of larger beetles for individual tracking or recapture to evaluate dispersal has led to the development of microtransmitters for radiotelemetry, and use of harmonic radar tracking. Pioneering studies using telemetry on the flightless *Carabus coriaceus* (Riecken & Raths 1996) have been followed by several studies on beetles of conservation importance. For example, movements of *Carabus auronitens* in Germany were studied by harmonic radar (Weber & Heimbach 2001). Studying migratory movements of the stag beetle *Lucanus cervus*, using transmitters cemented to the pronotum of beetles, helped to estimate connectivity between neighbouring populations in Germany. The greatest single flight distance recorded was 1720 m, with implications that females fly less strongly then males. Rink and Sinsch (2007) estimated that males may be capable of maintaining genetic interchange among breeding sites within a radius of up to about 3 km, but the lower dispersal capability of females (generally less than 1 km) suggested that populations more isolated than this may have increased likelihood of extinction. Males of the North American burying beetle *Nicrophorus americanus* can travel up to 6.1 km (one individual; Bedick *et al.* 1999), although most of the 158 marked individuals recaptured were much closer (92% within 1 km, 85% within 0.5 km). For *O. eremita* (above), radiotracking of beetles with small (0.48–0.52 g) radiotransmitters glued to their pronotum led to detection of transmissons up to 330 m, with 50–100 m routinely reported (Hedin & Ranius 2002); these transmitters appeared not to impede the flight of beetles in any significant way. As another example, radiotelemetry of the dynastine scarab *Scapanes australis australis* in Papua New Guinea was used to track individual beetle movements over periods up to 6 days. Some of these beetles were also marked with chemical light tags for short-term observations (Beaudoin-Ollivier *et al.* 2003).

The flightless grass-feeding cerambycid *Dorcadion fuliginator* has declined substantially in Europe and is now endangered, and confined to small remnants of dry grassland, having been affected by loss of habitat (Baur *et al.* 2002). Its preferred host plant is *Bromus erectus*, and larvae feed on grass roots for up to about 14.5 months. Adults emerge in midsummer but do not become mature until after a second hibernation period within the generation. All localities around Basel that currently support populations of *Dorcadion* are secondary, anthropogenic habitats with ruderal plant species. Dispersal occurs by walking, and mark–release–recapture surveys by Baur *et al.* (2005) showed a maximum movement of 218 m (by a single male over 12 days). The few beetles seen migrating walked along the linear verges of tracks or roads (apparently using these as corridors), and did not enter forests or arable fields. Field surveys and simulation studies suggested that beetles might effectively move only between populations separated by about 100 m. In contrast, populations or habitat patches separated by more than 500 m were functionally isolated. Extent of movement between patches decreased with increasing distance between them and, in Baur *et al.*'s study, most of the suitable patches apparently supported very low populations of *Dorcadion*. Although daily movements of this cerambycid were of similar magnitude to those of some large flightless carabids, the latter live for considerably longer so may have greater dispersal opportunity.

Some such beetles occur, whether naturally or as a result of habitat loss, in very small and highly localized populations. There is no universal rule over what constitutes a minimum viable population for a beetle and it is almost always uncertain as to what constitutes an effective population; however, in general, effective population sizes appear commonly to be only around 10%, at most, of a census population size (see Frankham *et al.* 2002). Such lack adds to the wider difficulties of appraising conservation status in terms of population size and dynamics. Pitfall trap samples and mark–release–recapture studies over two seasons helped to clarify activity, dispersal and population sizes of the long-lived (2–3 years) *Carabus variolosus* in Germany (Matern *et al.* 2008). At the two sites studied, the stream-bank habitat supported densities of around 0.85 beetles per 10 m^2 (site A) and 1.75 beetles per 10 m^2 (site B), giving estimated total beetle populations of around 215 (site A) and 150 (site B). These two sites were separated by about 2.5 km, and no marked beetle from either site was recaptured at the other. Matern *et al.* (2008) summarized data on other species of *Carabus* to reveal population densities well below 1 per 10 m^2 for seven other species assessed. As well as such low densities, however, the studied populations of *C. variolosus* appear to be smaller than might be viable in the long term, and may be prone to extinction simply because of their apparently marginal size, with risk perhaps enhanced by lack of dispersal in this flightless species. An important practical matter from Matern *et al.*'s study is that caution is necessary in exploring for other populations of this species, for which methods such as wet pitfall traps should be avoided because any additional mortality may be critical to such tiny populations. Teneral individuals should also not be captured, as they may be injured easily. Spring is therefore the preferred time for seeking additional populations, because such vulnerable stages may then not be present.

One practical caveat for studies on changes in numbers and persistence of some beetles, such as larger carabids, is that their longevity may be considerable. Numbers in successive seasons may therefore represent carry-over as well as recruitment of new individuals, sometime also confounded by movements. Marking of several Carabidae species in Europe has confirmed that they may indeed live for up to about 8 years; for *Carabus auronitens*, the oldest females marked by Weber and Heimbach (2001) reached their sixth season, and many of both sexes were recaptured in several seasons.

Beetle assemblages for conservation

Whilst individual species may respond to environmental change by decreasing or increasing in abundance or expanding or contracting in distribution, every beetle species is a member of an assemblage in which the different members are likely to be affected by change in different ways. However, the association of species in particular environments may be sufficiently stable to be regarded as characteristic (i.e. typical or representative) or indicative of a given set of conditions. Again, from Hanski's (1982) point of view, these comprise a suite of core taxa whose presence collectively diagnose the assemblage and, in turn, reflect the well-being of the environment in which it occurs. Changes in assemblages are thus the higher level of conservation concern, with changes in richness, relative abundance and composition all used as signals of disturbance and possible threat, not least to the assemblage's ecological capability in contributing to community-level ecological processes in the local environment.

At this level, several of the contexts mentioned earlier, and discussed also in later sections, have become predominant themes in advancing beetle conservation. Three terrestrial contexts, in particular, have been studied widely, and dominate much of the background development that supports wider conservation endeavour. Each field has also generated a vast body of scientific literature.

1   The assemblage of saproxylic beetles, particularly well studied in northern temperate forests and assessed both through focal species and by wider guild or richness interpretations, act as surrogates for wider forest health, advocates for the ecological values of dead wood, and standards by which to adjudge and ameliorate the impacts of forest management practices.
2   The assemblages of ground beetles in forest, urban and agricultural environments, and the factors that affect their change and resilience across the landscape and which can increase their values as generalist predators in agricultural pest management.
3   The mechanisms by which beetles track patchily distributed and ephemeral resources in the landscape, and the influence of those resources on assemblage structure and successional and seasonal changes. Dung beetles and carrion beetles have been predominant foci, but plant-feeding beetles are also represented.

Study of each of these lends itself well to experimental investigation, replication and beetle sampling by methods that, at least at first appearance, are simple and straightforward to standardize. Similar guilds and contexts can thus be studied comparatively in various parts of the world, and the baselines of species richnesss and assemblage composition integrated as templates against which to measure geographical variations and the effects of environmental change or conservation management. Studies on individual species can both draw on and contribute to this wider context, reflecting that the assemblage and community is the context in which any species must live and either thrive or decline.

A number of the examples of beetle richness in assemblages or wider regional samples noted earlier illustrate the difficulties of interpreting assemblages in functional terms, because the ecological relationships between species and their levels of specialization, as factors affecting their stability or vulnerability, will be poorly known or categorized simply by extrapolation from related taxa. A further practical difficulty, particularly in the tropics or remote areas of considerable conservation interest, is that assemblage changes may be revealed most reliably only by long-term studies, whereas some short-term changes may simply represent 'blips' following a disturbance or change. Most conservation planning does not, and perhaps logistically cannot, provide for monitoring and survey over a decade or more to determine whether such changes occur, despite repeated calls from commentators that such longer-term appraisals are necessary. Relationships between beetle assemblage composition and particular habitat features are often very difficult to determine but, at least in some studies, different functional groups (usually trophic groups) may differ in their responses. In woodland near Sydney, New South Wales, pitfall and flight intercept traps were compared in high-complexity and low-complexity habitats (Lassau *et al*. 2005). Although some general correlations were found between species richness and single environmental parameters (tree canopy cover, ground herb cover, amount of leaf litter, soil moisture), autocorrelations among these factors made interpretations difficult. Focusing on staphylinid subfamilies as functional groups, some more unambiguous responses emerged. Thus, Oxytelinae (detritivores) preferred habitats with more leaf litter, and Scaphidiinae (fungivores) occurred mainly in areas with high ground cover. In contrast, five predatory subfamilies showed no clear habitat preference, possibly because they forage widely.

Recognition and categorization of beetle assemblages depends on parallel robust recognition of biotopes, and raises the paradox that assemblages may be used to characterize biotopes and biotopes to define the boundaries of expected resident assemblages. As Eyre (2006) noted, massively increased beetle recording and survey effort in Europe has led to widely accepted schemes (see Chapter 1) whereby biotopes are defined on records of species assemblages, as revealed by multivariate analysis techniques. Eyre advocated the development of a 'single philosophy' that could be applied to explain distribution of beetle species, assemblages and biotopes, and drew on the considerable accumulation of data on ground beetles and water beetles in England and Scotland to illustrate how this might be attempted, incorporating aspects of productivity and disturbance as important influences. Their relevance is summarized for two examples from north-eastern England (Table 2.5) to indicate the principles involved. Grassland (ground beetles) productivity and disturbance was related to land cover and

Table 2.5 Biotope designations for grassland ground beetle biotopes and water beetle biotopes in north-east England, together with estimation of their productivity (P), disturbance (D) and substrate water (ground beetles) or water permanence (water beetles) (W).

Group	Biotope	P	D	W
Ground beetle grassland biotopes				
1	Upland grasslands and heaths	L	L	H
2	Coastal dunes	VL	VH	VL
3	Damp shaded grassland	M	M	M
4	Dry unmanaged grassland	M	M	L
5	Very dry open grassland	L	H	VL
6	Damp unmanaged grassland	M	L	M
7	Intensively managed lowland grassland	VH	VH	M
8	Lowland pasture	H	M	M
9	Very wet unmanaged grassland	L	L	VH
10	Managed upland grassland	H	H	H
Water beetle biotopes				
1	Fast flowing rivers and streams, bare	M	VH	H
2	Large open lakes, little vegetation	M	M	H
3	Lowland vegetated ponds	H	L	H
4	Transition mire, mid-altitude, vegetated	M	L	H
5	Marshes, dense vegetation	M	L	M
6	Lowland slow-flowing water, vegetated	H	M	M
7	Temporary lowland ponds, vegetated	H	L	L
8	Upland mires, dense vegetation	L	L	H
9	Upland slow-flowing water, vegetated	L	M	M

VL, very low; L, low; M, medium; H, high; VH, very high.
Source: Eyre (2006) with permission.

agricultural activity, with lowland grasslands more productive than upland ones because of peat-based soils in the uplands and reflected the importance of suitable water levels.

For water beetles, physical disturbance relates to water flow and wave action, and productivity reflects geographical location, with the most productive biotopes occurring in lowland areas. Eyre (2006) suggested that if accurate measurements of these parameters could be generated, they could be informative for conservation plans for single species but also help to change focus for greater collective benefit by concentrating on assemblages and biotopes. The ability to achieve this depends on use of standardized sampling and recording schemes.

3

Threats to Beetles: the Role of Habitat

The variety of threats to beetles is almost as broad as the variety of the insects themselves, but two very broad categories dominate conservation interests. These devolve on changes to habitats (equated largely to biotopes) from human activity and accusations of over-collecting, but an array of other events and processes, many associated with effects of pollution or invasive species, are nominated and implicated more sporadically. Overriding any more local issues, the implications of current and future climate changes on distribution and well-being of beetles are less tangible but possibly severe and, in many cases, inevitable. Many beetles occupy habitats that we may regard as in some way extreme (highly saline marshes, arid deserts) and, as for specialists in any other regime, can become vulnerable to even moderate changes.

Many of the environmental changes discussed here as threats are by no means only recent occurrences or trends. Thus, Whitehouse (2006) noted that up to 40 beetle species in Great Britain and 15 in Ireland had become extirpated in the Holocene, as inferred from their disappearance from the fossil record. Some of the Irish species, in particular, were lost only a few hundred years ago, during a period when the landscape was being extensively cleared. Purported reasons for these extirpations (collectively evident from about 4000 years ago) include a combination of forest clearance and human activities, as well as the consequences of climate change on distributions (see p. 133). High proportions of the extirpated species are saproxylic, with changes to the natural disturbance regimes that ensured the continuing mosaic supply of their critical resource (namely old trees and dead timber) being implicated (Kaila *et al.* 1997). Consequently, the more specialized resources, such as different species of *Polyporus* fungi, needed by particular beetles were also lost or dispersed more widely, so that their patterns of availability were changed. Forest history, as for modern beetles, may have

Beetles in Conservation, 1st edition. By T.R. New. Published 2010 by Blackwell Publishing.

been a decisive influence on beetle well-being, either positively or negatively, for at least several thousand years. Many of the extirpated species have low mobility (Whitehouse 2006) so that forest fragmentation through deforestation may have had severe effects. More limited factors, such as declines of conifers, may also have contributed to losses, with nine of the 40 species extirpated in Great Britain associated with pines. Structural changes in forests, in some cases linked with compositional change, may alter fire regimes and canopy cover – the openness of the forest – and any of these changes might contribute directly or indirectly to the pool of threats increasing the vulnerability of beetles and other organisms. Conversely, any such change may be a benefit. Thus, gap size in boreal forest management in North America influences the composition of staphylinid and carabid assemblages differently, reflecting that a high proportion of Staphylinidae there (pool of 116 species) are forest specialists and cannot adapt to large gaps, whereas more of the carabids (pool of 38 species) tolerate open ground easily. Forest gaps may thereby influence different group of beetles in different ways (Klimaszewski *et al.* 2005, 2008).

Determining the most important threats to any focal species or assemblage is central to planning constructive conservation management, and broad generalities must then be considered in the more specific biological contexts that prevail. An initial caveat, sometimes not emphasized sufficiently but evident in many of the examples that follow, is that the impact of any threat or environmental change is context-influenced or context-dependent, so that the impact of a given disturbance in one place or assemblage may not be paralleled strictly in another. Logging causes less disruption to boreal forest carabid asssemblages in Finland than in Canada (Niemela *et al.* 1994), possibly because forestry impacts have occurred for far longer in Finland so that generalist beetle species able to withstand some level of change are distributed widely. Canadian forests, then being felled for the first time, still harboured specialized forest species not well adapted to cleared areas. Changes of tree species, such as the adoption of plantation forestry where monocultures replace multispecies stands, may strongly influence assemblage composition, mainly at the expense of species that cannot utilize the new tree species. Even when native tree species are involved, changes in species richness can occur: thus in Japan, differences in Curculionidae (Ohsawa 2005), Elateridae (Ohsawa 2004a) and Cerambycidae (Ohsawa 2004b) occurred between a plantation of Japanese larch (*Larix kaemferi*) and broadleaved forests (see also p. 193).

Each such threat or environmental change may lead to (i) local or more widespread loss or (ii) decline in abundance and/or distribution on scales ranging from single population or site to national or global. Detecting declines of individual beetle species or their richness in assemblages is often difficult, with apparent trends sometimes reflecting normal background variations in numbers. The major exception is the obvious alienation of habitat through imposed changes. Otherwise, changes may be gradual and insidious and, in most cases, rigid data on beetle richness and abundance before, during and after the changes will not be available and any information available is likely to be inferential and not collected by defined sampling effort or formal plan to test changes due to the impact.

Some specific threats to individual taxa may not initially be obvious: vehicles crushing beetles on roads, collecting for the pet trade, or even the name of a beetle becoming a supernormal inducement for collectors are examples. The first is regarded as a serious threat to the flightless dung beetle *Circellium bacchus* in the Addo Elephant National Park (South Africa), where visitors are greeted at the park entrance with signs proclaiming 'Dung beetles have right-of-way', repeated at intervals along the park roads. The second is exemplified by the likely threat to South-east Asian stag beetles from being adopted as pets in Japan (see p. 118); and the last has recently received publicity in the context of a European carabid beetle named after Adolf Hitler (Elkins 2006). This rare beetle occurs only in a few caves in Slovenia, and the attraction of the specific name of *Anophthalmus hitleri* to right-wing zealots and to entrepreneurs catering to the lucrative Nazi memorabilia market has led to it becoming a cult object. Despite formal legal protection in Slovenia banning all collecting, the proximity of sites to the Italian border has reportedly reduced the effectiveness of this protection. In another context that can cause concerns, one threatened beetle species may pose a threat to another. The threatened European click beetle *Elater ferrugineus* in old hollow trees has predatory larvae that feed on larvae of other saproxylic insects in the hollows, including those of *Osmoderma eremita* (see p. 84). The adult click beetles use the male-produced sex pheromone of *Osmoderma* as a kairomone to help locate their prey. It has been suggested that male *Elater* also use it to help find mates and locate suitable breeding hollows (Svensson *et al.* 2004). On the one hand, the pheromone may be a useful monitoring tool for use in assessing both these beetles (see p. 25); on the other, it may be instrumental in increasing predation on the larvae of *Osmoderma*.

The destruction of habitats, or less obvious changes resulting from human activity, are a universal concern in insect conservation. Many such habitats have always been restricted in extent: the loss of coastal beach habitats to human recreation and related developments, for example, may be important for any specialized beetle. Several cases of shoreline tiger beetles are noted in this book, and another example is of the British staphylinid *Scopaeus minimus*. Lack of any post-1969 records in southern England may be due to loss of its coastal shoreline habitats (Hammond 2000). Ecologically specialized taxa, be they beetles or others, are widely susceptible to changing exposure or floral composition, incidence of specific plants, changes in water quality, reduction and fragmentation of suitable habitat patches and their connectivity in the wider landscape, supply of any critical resource, and microclimates, among many other factors. Localized distribution, whether geographical or ecological, may contribute to vulnerability, but establishing details of the less obvious influences of changing resource balance is often difficult. Collectively, beetles are sufficiently diverse that almost any projected or actual change or local development may be accused of posing a threat to one or more localized species, and one of the practical problems is simply to assess which cases are truly of concern. Some trends of change afford greater consensus: for example the dependence of numerous beetles on dead wood has led to calls to conserve this as a critical resource in forests (Speight 1989; Grove 2002) and to ensure its continued supply rather than remove it for sanitation or firewood or to decrease severity of future wildfires. More ambivalent are changes such as construction of roadways through forests, as they may constitute barriers to some species (Mader 1984) but corridors for movement of others. Roadside and adjacent forest assemblages may differ considerably, with the former having greater representation of open-ground species (Koivula 2005), some representing immigration along the road. A 3-m wide dirt road almost halved the number of Carabidae moving from one side to the other and so had considerable influence on population integrity (Duelli *et al.* 1990). Assessment of habitat suitability and quality for beetles (and other insects) is intrinsically interwoven with landscape ecology. For example, barriers may be formed by relatively small disturbances that may not initially be considered significant but which functionally impose isolation and increased fragmentation of populations or assemblages (see p. 68) by curtailing natural dispersal. Mader's (1984) study involved marking individuals of *Abax parallelepipedus* (Carabidae) and releasing them on road edges. Only one of 742 beetles marked was recorded crossing a highway lane. Mark–release–recapture studies by Koivula and Vermuelen (2005) included 10 species of Carabidae, nine of which were forest species and did not cross roads, although some moved to the road verge. Only one species, the open-habitat *Poecilus versicolor*, was found to cross the road (with 22 of 225 marked individuals doing so). However, an earlier study of the open-habitat species suggested that road crossings are relatively rare: 121 of 1475 marked beetles (*Poecilus lepidus* 200, *Harpalus servus* 1000, *Cymindis macularis* 275) were captured, but none had crossed a road. Such apparently simple habitat divides as roads can influence behaviour of different beetle species in different ways. Any such area increases exposure to predators and to being crushed by traffic. However, although a road was clearly a barrier to two species of carabids tested in Finland, the reaction of the beetles to encountering a road differed. *Poecilus versicolor* turned back and re-entered the patch, whereas *Agonum sexpunctatum* moved along the road verge (Noordijk *et al.* 2006) so that the barrier posed different redispersal consequences for these species. Koivula and Vermuelen (2005) inferred the likely high significance of road construction in Finland and the Netherlands in affecting spatial structure of beetle (and other invertebrate) populations. Dispersal capability varies greatly among beetles, and some carabids may be able to fly for only part of their adult life (see p. 176) or not at all.

Yet clarifying the impacts and roles of even simple interruptions to continuous habitat on beetles is extraordinarily difficult. Roads through forests, for example, come in many forms – paved or unpaved, narrow or wide, and with varying forms of edge – and may affect species in at least three different ways: (i) reducing available habitat, (ii) affecting patterns of movement and (iii) extending edge effects into forests (Dunn & Danoff-Burg 2007). For carrion beetles in New York, richness was lower near highways and two-lane paved roads relative to that near dirt roads, with 80% of the species near dirt roads lost near highways. The causes are difficult to determine, but Dunn and Danoff-Burg noted that more compacted soils may prevent the beetles burying carrion, and that the supply of suitable carrion itself may vary, both in amount and in competition for it; or that microclimate differences such as those related to litter or vegetation cover might be involved. Staphylinidae increased near highways, but the major inferences from this study included that high-use paved roads might have greater effects than smaller paved roads on carrion beetles.

Habitat continuity and the simple availability of the critical resources and conditions suitable for a given beetle species are intricately related.

Habitats

Understanding the relationships between a species or assemblage and its habitat is the greatest and most universal need in practical conservation management. Vast numbers of studies relate distributions and abundances of beetles to features of their habitats. Very broadly, these studies fall into about four main categories and, although undertaken for a variety of purposes, may contain information of relevance for conservation. In some, this is a direct purpose of the work, but most contribute to basic biological understanding of the species or assemblage involved, or indicate features of biomes or resources that merit attention in conservation of similar taxa or environments. The major categories include the following.

1 Autecological studies in which the needs of individual focal species are emphasized, leading to clarification of phenology, population dynamics and resource needs, and sometimes also the influence of changing habitat features.
2 Broader taxonomic-based studies, in which related species, usually at the family level (with Carabidae a predominant focus) but sometimes with greater emphasis on guilds, are surveyed and their relative abundance and seasonal incidence assessed in one or more broader biological contexts.
3 This approach is sometimes extended to emphasize beetles in disturbed habitats, such as managed forests and agricultural systems, with inferences that richness and abundance are affected by disturbance and, if so, to clarify what those influences may be.
4 Comparison of beetle assemblages in a series of different habitats, such as vegetation types or water bodies, usually within the same region, and sometimes related to successional changes or management regimes.

Habitat changes vary considerably in intensity and extent, from entire loss over large areas to relatively small diminution in quality influencing carrying capacity rather than occupancy, and over only small areas. Entire assemblages or one or a few species (or some intermediate quantity) may be affected, and many changes lead to increased abundance of some resident species or facilitate colonization by species not present previously. The natural hierarchy of habitats, illustrated effectively by Hanski's (2005) simile of 'matrioschka dolls', is a necessary consideration for beetles, as for other insects. In Hanski's example, the habitat of the endangered Finnish *Pytho kolwensis* (Pythidae), the larvae of which feed on phloem of spruce trees, includes the following levels:

Boreal forest
   Spruce-dominated forest
      Spruce mire forest with high temporal continuity of fallen logs
         A fallen spruce log with the base above the ground
            A particular stage in the decay succession of phloem under the detatching bark

These stages, once recognized, do much to characterize the requirements of any focal species in terms of ecological need, but a somewhat different parallel hierarchy relates to spatial scale. Thus the abundance of the fungivorous *Bolitophagus reticulatus* (Tenebrionidae), a species that is monophagous on the bracket fungus *Fomes fomentarius*, was analysed at four spatial scales (individual fungus, tree, tree-group, forest island) to reveal different patterns of beetle emergence at these different levels (Rukke & Midtgaard 1998).

The need to define critical resources in the context of the major habitat for any species is, as emphasized by Dennis *et al.* (2006, 2007), an avenue towards defining a species' habitat in terms that are meaningful for conservation, and moving away from the more embracing attributes of 'place' alone. Evaluation of the critical resources needed by a species as two main groups (consumables, utilities) helps to pinpoint the needs of the species for conservation within the wider habitat context. However, as also emphasized by Dennis *et al.* (2007), in these terms the habitat of only very few species can be characterized fully, with most conservation attention devolving on wider features such as major vegetation type (biotope) and the characters of suitable sites projected and defined largely on correlation of the species' presence (and, more rarely, abundance) with such broad features. One example cited by Dennis *et al.* (2007) is of the ground beetle *Carabus intricatus*, a red-listed species in Britain, where it occurs only in (i) deciduous woodlands in (ii) steep-sided valleys, predominantly (iii) running east–west, (iv) containing flowing water, (v) having an annual rainfall greater then 150 cm, (vi) atypically grazed by sheep and (vii) occupied by the slug *Limax marginatus*. The combination of the seven features noted above appears to be very precise and definitive information, but there are many unknowns. As Dennis *et al.* noted, the cited study by Boyce and Walters (2001) has identified surrogate markers for resources and environmental conditions suitable for the beetle and used these to design management. This example is paralleled by numerous other insect conservation programmes in that the fine details of resource needs and uses remain unidentified, or the reasons for the restriction are not understood. In general, the habitats of beetles defined in conservation programmes are more akin to biotopes, so that much conservation has focused primarily on species or assemblages in such situations as grassland, heathland, riverine sediments, lakes, forests and so on, with varying levels of finer-level subgroupings within these. In general, these broad units parallel characterizations cited in general reference works, many of which also contain considerably more information on specific plant or prey associations. However, particularly for rare species, it is easy to be misled from either field observations or literature records. In their compilation of the host-plants of the world's Chrysomelidae, for example, Jolivet and Hawkeswood (1995) had to discard a substantial number of published records because of uncertainty or unreliable identification. And just because a beetle is found on a given plant does not necessarily mean that it feeds on it or, even if feeding traces are found, that it can thrive there. Many plant-feeding insects will test plants on which they may be casually present and if unable to disperse further may be committed to staying there. Jolivet and Hawkeswood estimated that reliable food plant records are available for around 30% of their estimated 35,000 chrysomelid species, but many of those records are fortuitous and, however valuable, may not represent the full spectrum of food plants taken. As they note, 'there are many gaps in knowledge to be filled', and the comment applies to resource needs of many different feeding guilds of beetles.

The great variety of beetle biotopes, and the inconsistency or imprecision of terms used to characterize them in relation to the assemblages they support, has led to difficulties in comparing different studies and contexts, and to calls for more effective standardization. Eyre (2006) emphasized the need for better biotope classification and attempted to produce a system for beetles based on the twin axes of productivity and disturbance. For grassland Carabidae, for example, productivity is related to soil quality, whilst disturbance relates more to cover and/or land management. Using water beetles and ground beetles, both very well recorded in Britain (see p. 40), Eyre showed that interpreting records from the 10×10 km recording units in these terms might be useful, notwithstanding the additional information that could be added by considering other parameters as well, and could usefully encourage further refinement of their use. Productivity and disturbance were reflected in traits of species constituting the assemblages: disturbance was associated with different morphological traits in water beetle species of lentic and lotic waters and in ground beetles at the two extremes of land management (Ribera *et al.* 2001). Land management was adopted as a surrogate for land disturbance, and habitat adversity or stress measured by elevation and vegetation structure for Carabidae in Scotland, with the ensuing range of environmental variables summarized in Table 3.1 (Ribera *et al.* 2001). The 68 carabid species showed strong correlations between morphological and life-history traits and the main environmental gradient of their habitats, so that functional groups could be designated. For example, plant-eaters were associated mainly with highly disturbed sites (and distinguished in part by having long trochanters and wide femora), while Collembola-feeders were associated with agricultural fields where springtails are abundant as prey (and the beetles distinguished in part by being diurnal and with metallic body and legs). In contrast, black and pale-bodied species tend to be nocturnal, with many pale species burrowing or living on bare ground. A high proportion of this group may be found on disturbed agricultural sites.

Continuity of suitable resources, in ways in which they are in sufficient supply and are accessible to the beetles needing them, is the basis of informed management for habitat suitability. A consequence of the common lack of detailed knowledge of resource requirements and of functional correlations with beetles is that, whilst good comparative studies of beetle assemblages can indeed quantify differences in species composition and incidence, the precise causes of such differences are necessarily largely inferred from correlations with

Table 3.1 Land-use variables and states calculated to constitute a management intensity score to help characterize environmental influences on carabid assemblages at a range of sites in Scotland. Each variable was scored from 0 (none) to 3 (high) in ascending order of intensity of management.

Variable	Attributes and score categories (0–3)
Sward type	Natural or semi-natural Sown or improved Grass mixture Ryegrass ley
Age of current land use	Uncultivated > 10 years 5–10 years < 5 years
Soil disturbance	None Only harrowed in last 3 years Ploughed once in last 3 years Ploughed twice or more in last 3 years
Cutting	None Topping only One complete cut and removal of vegetation Two or more complete cuts and removal of vegetation
Grazing	Scored on basis of number of livestock, if any, from < 0.8 to > 1.14 LSU/ha
Inorganic fertilizer	None NPK < 50 kg/ha NPK 50–100 kg/ha NPK > 100 kg/ha
Organic fertilizer	Four levels, from none to heavy manure dressing
Pesticides	None Fungicide only One herbicide and/or one fungicide Two or more herbicide products and/or insecticide and/or glyphosate

LSU, livestock units; NPK, nitrogen/phosphate/potassium.
Source: after Ribera *et al.* (2001) with permission.

environmental factors, or by analogy with studies elsewhere and in different environments. More generalized threats to assemblages may also occur at quite coarse scales or be much more taxon-specific in their impacts. Site-specific factors may have considerable influence, related both to the variety of possible resources present and to the particular variables important for focal beetle species or larger components of assemblages. Habitat structure in terrestrial biomes is most commonly expressed in terms of vegetation form or composition, and this alone may not relate to major differences in some beetle families. Passalidae in primary and secondary forest in Mexico differ rather little (Castillo & Lobo 2004), with only two of the 12 species preferring one or other forest type. In this example, perhaps exceptionally, disturbance of primary forest may have little effect on assemblage composition. Passalidae depend on decaying trees (whether standing or fallen), a resource less abundant in secondary forests so that lack of difference from primary forest initially appears contradictory (Castillo & Lobo 2004). Possible additional factors might include (i) less competition between species in secondary forests and logging helping to reduce the dominance of more abundant species there, and (ii) that the tree species is of little importance to the beetles.

It is almost inevitable that samples from beetle assemblages in any particular habitat will include species present casually or which are not confined to that habitat and so not fundamentally dependent on it. Dependence of a beetle on a specific habitat is often difficult to prove but, following the example discussed by Sadler *et al.* (2004), it may be feasible to derive some index of fidelity to a biotope or resource from collection data and background knowledge. Using the British beetles associated with exposed riverine sediments (ERS), Sadler *et al.* recognized two such categories, based on ecological attributes of the beetles themselves rather than simply presence or absence alone. Thus, grade 1 species depend for at least some stage of their life cycle on bare or sparsely vegetated sediments on river banks, although some may occur also on exposed lacustrine sediments. Grade 2 species are strongly associated with ERS for at least some stage of their life cycle, but are also characteristically found in other habitat types with extensive deposits of bare sediments, such as on sand dunes or in sand or gravel pits. Intuitively, such indices reflect extent of ecological specialization and, as for numerous other studies of beetles in many different terrestrial or freshwater biotopes, that specialization may equate to vulnerability to imposed change. The resources involved indeed encompass utilities, either including microclimate or (following a number of recent treatments) categorizing this separately as a conditioner. In the above example, hibernation potential was separated into three categories on site variables involving (i) amount and diversity of buffer habitat in the river corridor, mainly riparian vegetation, tussocks and dead wood; (ii) diversity and nature of vegetation and dead wood on the sediment areas themselves; and (iii) the nature of the substrate, reflecting that smaller beetles may overwinter in the sediments and so require sandier or more open gravel sediments to furnish suitable sites.

Assemblage structure and richness may be difficult to define, as a needed template against which to assess changes. Several recent studies of tropical forest canopy Coleoptera illustrate some of the points and problems that need to be considered. Beetles are abundant there at any time of the year, but both richness and the relative representation of major taxonomic components may differ substantially across seasons. In Uganda, canopy fogging of two tree species revealed strong beetle species overlap between tree species within a season, but much less overlap between conspecific tree individuals between seasons (Wagner 2003). The greatest decrease in abundance from wet to dry season was in the mycetophagous beetles. One interpretative problem in studying such diverse assemblages, noted by Floren and Linsenmaier (2003), is simply defining the

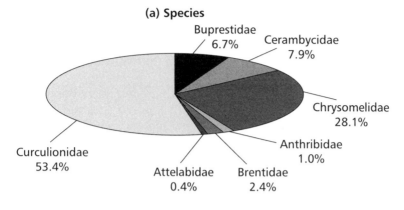

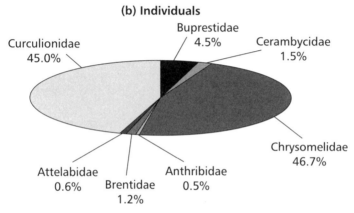

Fig. 3.1   Beetle richness: the proportions of phytophagous beetles in predominant families enumerated by (a) species and (b) individuals collected from forest canopy in Panama. (From Odegaard 2003 with permission.)

assemblages properly, because they may not be distinguishable from randomly composed sets of the constituent beetle species. Thus, any tree-specific (or, by extrapolation, other resource-specific) assemblage cannot be identified or characterized easily. This is in marked contrast to beetles of temperate forests, where such specificity has become largely a well-documented aspect of beetle natural history. Nevertheless, differences are sometimes definable adequately in the tropics, based on the approach that relative numbers of phytophagous and xylophagous beetles tend to be highest on their preferred host-plant species. One study towards this end in Panama (Odegaard 2003) involved analysis of 1165 beetle species collectively represented by more than 35,000 individuals from tropical dry forest canopy incorporating 50 plant species with each individual plant sampled weekly over a year. The proportions of major taxa of phytophagous beetles (Fig. 3.1) were highest for Curculionidae (616 species) and Chrysomelidae (300 species), but the overall inference was that 7–10% of all species were monophagous, a figure substantially less than Erwin's (1982) suggestion of 20%. Flower-visiting tropical beetles may also show substantial segregation, but Kirmse *et al.* (2003) suggested that considerable stochasticity may be involved in determining the variety of beetles found on any plant species. In Allison *et al.*'s (1997) Papua New Guinea survey (see p. 31), only five families (of 53 present) yielded more than 200 individuals each, in contrast to the numerous low abundance taxa collected.

The ecological diversity of beetles ensures that different resource components may be particularly important to individual taxa on a site, and may be related to management of the site, perhaps superimposed specifically on broader, more general resource conditions. In a European beech forest, as one example, Müller *et al.* (2008) showed the following broad correlations: (i) the amount of dead wood was important for Elateridae and overall beetle richness; (ii) the amount of flowering plants affected the richness of Cerambycidae; and (iii) that of wood-inhabiting fungi influenced richness of Staphylinidae and overall beetle richness. The frequency of the bracket fungus *Fomes fomentarius* was linked to richness of threatened beetle species. At either species or larger group level, habitat specificity or broader preference by beetles often reflects foraging requirements and, more specifically, food needs in combination with microhabitat as influenced by the complexity of the local environment. In a related but taxonomically wider example, of saproxylic beetles and hoverflies in Belgium, Cerambycidae favoured oak-dominated stands with a high volume of snags, whereas hoverflies (Syrphidae) were restricted to more open stands in which a rich floral herb layer could develop as a nectar supply (Fayt *et al.* 2006). Within a forest near Sydney, Australia, Lassau *et al.* (2005) found a positive correlation between beetle species richness and each of tree canopy cover, ground herb cover, amount of leaf litter, and soil moisture. Although separation of such factors as individual components of the environment is overly simplistic (as acknowledged by these authors), particular guilds of beetles could be correlated more logically. Thus, guilds of Staphylinidae present included detritivorous Oxytelinae (preferring habitats with more leaf litter) and fungus-feeding Scaphidiinae (most abundant in sites with substantial ground herb cover), whereas predatory subfamilies showed no such clear preference for any of these variables – as active predators they may naturally forage much more widely for their prey. Resource specificity for beetles extends well beyond simply the species of plant or prey used for food, to include much more fine-grained requirements. For example, in a survey of a variety of variables of 145 sample spruce trees, the incidence of *Pytho kolwensis* (above) was correlated positively with diameter of the host logs (Siitonen & Saaristo 2000). Although beetle larvae were found in many kinds of trunks, 87% of the 55 inhabited trunks shared the following features: (i) trunk diameter at breast height at least 20 cm; (ii) bark cover at east 50%; (iii) average penetration by a knife blade at most 1.5 cm; (iv) mycelium cover of at most 75%; and (v) trunk lacking contiguous contact with the ground. Various overlapping inferences on causes of such preferences involve microclimate, host tree quality, competition and interactions with other species, stand structure, and habitat continuity. Siitonen and Saaristo (2000) favoured the last of these as the most important determinant.

Beetles are abundant and diverse in dead wood, which supports some of the richest beetle communities found in forests and is a major consideration for conservation; they are discussed in many contexts in this book. Beetles are commonly among the pioneers of complex saproxylic communities, and some are among the first insects attracted to volatile chemicals emitted by weakened, stressed or newly dead trees. Commonly, these early colonizers, which may arrive by flying or walking so that fallen logs and standing trees may garner different communities (Hammond *et al.* 2001 on *Populus tremuloides* in Canada), are dominated initially by species feeding on fresh phloem and wood. These diminish after a few years and are replaced by a succession of other species as the wood ages and decays, perhaps over many decades. In boreal spruce forest, a high proportion of beetle species (e.g. 232 of the 553 species collected by Martikainen *et al.* 2000) are associated with dead wood. Similar proportions have been recorded elsewhere: Kohler (2000) noted that 56% of all forest beetles in the northern Rhineland of Germany are saproxylic. Not altogether unexpectedly, the richness of these saproxylic species is greater in old-growth forests than in managed forests, and in Finland many (78%) were more abundant in old-growth than in mature managed forests; the assemblages in the two forest categories overlapped very little. In contrast, richness and abundance of other beetles differed relatively little, so that these may require far less attention to their conservation in concert with forest management. The key need in this system seems to be to conserve the small remaining proportion of old-growth forest, because even managed forests left entire well beyond the usual rotation period of 90–100 years do not fully attain the fauna found in old-growth forest.

Grove (2000a) tabulated a selection of correlative or predicted relationships between saproxylic beetles and diameter of wood attacked, and found similar patterns across studies from various parts of the world. However, whilst particular beetle species may show very specific trends or preferences, closely related species may differ considerably. For example, species of *Agrilus* (Buprestidae) in North America include those with rather precise branch diameter needs and others that are much less selective (Hespenheide 1976), with the size of the beetle correlated with the size of the branch attacked. The Japanese lucanid *Prismognathus acuticollis* occurs more usually in smaller-diameter logs (Araya 1993). Grove's (2000b) studies across 81 sites in tropical Australia (and including 118 saproxylic beetle species) confirmed the wider impression that species richness tends to be higher on larger-diameter trees and dead wood, so that large trees can be particularly important for conservation of many rarer beetle species. The values of old-growth forests, noted above, rest in part on the lack of vagility of many such species, so that continuity of resources present is critical. Those resources have not yet been attained or paralleled fully by accumulation during succession following forest exploitation and management. As Grove (2002, following similar sentiments by Speight 1989) put it, in referring to large-diameter veteran trees 'They may be commercially overmature but are in their ecological prime of life' with both standing dead trees and fallen wood vital resources for numerous invertebrate species. Many of the saproxylic beetles of greatest conservation concern in Britain are associated with later stages of decay, particularly in such ancient veteran trees.

In a review of management for saproxylic invertebrates, Davies *et al.* (2008) stressed the considerable variety of manipulations that occur in silviculture and the corresponding variety of invertebrate responses to these, so rendering broad generalizations very difficult. Numerous gaps remain in basic ecological knowledge, and there is widespread lack of capability for long-term monitoring – most studies are relatively short term but describe changes in response to particular management practices. As others have also done, Davies *et al.* emphasized that many of the species involved and of greatest concern are specialists, so that generalized measures for conservation of the assemblages present may not cater individually for all the significant species. This principle is of much wider application, and endorses the wider difficulties of relying on putative umbrella species as independent conservation measures. The decay profile of wood, and the spectrum of beetles present, can differ according to numerous factors. Alexander (1999) showed that simply allowing the woody material to decay in sunny or shady conditions can influence the assemblages present.

Nevertheless, the major threat to saproxylic beetles is simply depletion, through deliberate removal, of dead wood from forests. Retention of this in silvicultural stands is now an integral part of good conservation management, whether standing or fallen wood. Ground debris is an important resource for many litter-frequenting invertebrates, and is associated with increased richness and abundance of fungus-feeding and predatory beetles in forest litter (Topp *et al.* 2006, for Slovakia), with these decreasing with distance from coarse litter. For managed forests, Topp *et al.* recommended increasing the amount of this litter left on the forest floor, both to ameliorate the environment and to help sustain nutrient recycling. Without coarse woody debris to provide shelter and food, populations of litter-dwelling beetles may become more vulnerable to local extinctions. Both the quantity and quality of such coarse woody debris are important in order to produce a heterogeneous substrate through a variety of conditions and management practices (McGeoch *et al.* 2007). Collectively in Fennoscandia, around 400 saproxylic beetle species have been considered to be threatened as a direct result of such resources being lost through forestry, with reduced size and quality of old and dead trees, logs and stumps affecting many species (Jonsell *et al.* 1998). Including beetles living in bark and wood, the Swedish fauna alone includes 1065 listed species, of which 446 are saproxylic (Jonsell *et al.* 1998) and these vary considerably in their specializations and preferences. Also for Sweden, Niemela (1997) estimated that more than half of the saproxylic species were sensitive to forest thinning and clearfelling. In such contexts, even individual trees occupied by notable beetles are significant conservation units and may be designated individually for protection. Trees supporting the endangered scarabaeoid *Osmoderma eremita* are regarded as important for conservation, and may also support a range of other red-listed or otherwise significant species. In a survey using a vacuum cleaner to extract debris from tree hollows in 117 oak and 10 beech trees in Bavaria, 17 of 35 beetle species recovered were in these categories (Bussler & Müller 2009), with *O. eremita* recorded in 39 trees, which were subsequently marked to aid their future individual recognition and protection during forest management.

Management practices such as burning of logged sites can have very mixed effects, and illustrate well the delicate balance between honed habitat management and threat in insect conservation (see p. 147). Species richness of saproxylic beetles in northern European forests can benefit from burning sites after logging, but that benefit may depend on retention of some trees. Differences between burned and unburned sites increase with numbers of trees retained (Toivanen & Koticho 2007), and the effects of burning are not significant unless about 15 trees/ha are kept. In particular, many rarer and red-listed species unlikely to persist in normally managed forests strongly benefit from this management combination. However, beetle richness eventually decreases over time, so that a continuum of burned areas is needed to assure their conservation. Burned logged sites may provide resources for saproxylic beetles for only 10–15 years, in contrast to areas suffering a natural fire, in which dead wood may persist for several decades.

Broadly, the effects of burning on beetles are difficult to predict or explain. Burned boreal forests in Europe support a characteristic insect fauna, but the kind of tree involved may lead to very different outcomes. Wikars (2002) compared the beetle fauna of burned and unburned logs of birch (*Betula pendula*) and Norway spruce (*Picea abies*) by collecting the insects emerging from netted logs for 2 years after tree death. The pool of beetles collected comprised 142 species and the assemblages responded to site differences as much as to differences between burned and unburned logs, but more than ten times the number of bark beetles (Scolytidae) emerged from spruce than from birch logs. This study demonstrated that certain species of beetles need burned trees, confirming that burned forest differs considerably as a habitat from unburned forest, so that forest patches may need to be burned regularly as part of the conservation of those species. Reasons for this preference are unclear, but two possibilities (Wikars 2002) are that the species include (i) mycophagous beetles that need burned substrates because the ascomycete fungi on which they depend favour burning, and (ii) phloem-feeders and predators are favoured by some unspecified characteristic(s) of recently burned forest. Some open-habitat beetles may benefit from presence of firebreaks in forests (Mawdsley 2007).

High levels of host-plant or other resource specificity are commonly inferred in beetle assemblages, as one factor contributing to high levels of species packing. However, monophagy or parallel specializations are difficult to prove, and many such inferences result from correlative surveys rather than experimental trials. An investigation of Cerambycidae in French Guiana (Tavakilian *et al.* 1997) involved rearing beetles from 690 individual felled trees collectively representing around 200 species of 38 plant families, and yielded 348 longicorn species. Guilds of species associated with different plant families had very different ratios of specialist to generalist species although, overall, specialists outnumbered generalists by more than 3 : 1. In this study, specialists were those reared from a limited range of (usually related) hosts. However, many species could not be categorized reliably, and interpreting their status raises a problem of much more frequent occurrence in beetle surveys. Tavakilian *et al.* noted that many species were represented by a single host-plant record, so that several possible inferences could be made: (i) they might indeed be rare specialists on the plants sampled; (ii) they might be specialists on plant species not included in the survey; (iii) they might reproduce mainly in living wood, or wood at a decomposition stage not included in the survey; and/or (iv) they might breed at a different time of year and so not be revealed in a seasonally limited survey.

It is not uncommon for our ideas on the habitat relationships and resource needs of beetles to change markedly when a species or assemblage is studied in greater detail. In saltmarshes, for example, the relatively few species of strict halobiont beetles co-occur with many other species that are salt-tolerant but drawn to those habitats by other factors such as microclimate and vegetational features, rather than by the saltmarsh environment itself (Foster 2000). Representatives of 10 genera of British Carabidae contain halobiont species, but only two small genera (*Dicheirotrichus*, *Pogonus*) are exclusively so (Luff & Eyre 2000). Such variety suggests that, as for water beetles in similar habitats, preference for the saltmarsh habitat has arisen independently on several occasions. As for phytophagous beetles limited to obligately saltmarsh plants, these may simply be isolated occurrences of habitat-bound specialization whilst most of their close relatives are still found elsewhere. As a third family example of species associated with saltmarshes, very few of the Staphylinidae found in intertidal regions are associated primarily with saltmarsh (Hammond 2000). With slight relaxation of criteria to accommodate burrowing species, several species of *Bledius*, for example, are 'more-or-less saltmarsh specialists' (Hammond 2000, p. 277). Other *Bledius*, deemed halobionts by association with these, may in reality be only fringe members of this specialized group. Moisture regimes, sand aeration and level of tidal inundation may all be important influences on surface or burrowing beetles in these environments. For phytophagous or predatory species, clarifying the existence and extent of host or prey specificity depends on such intensive rearing trials and accompanying screening tests. Nevertheless, relatively sound inferences can be possible from the evidence of regular incidence or relative abundance of beetles across possible hosts as an indication of fidelity (see above). Thus, Paulay (1985) categorized the 67 species of small flightless *Miocalles* weevils on the Pacific island of Rapa into four groups on the basis of where he found them:

1   very host-specific species, found only rarely wandering on plants other than their host;
2   moderately host-specific species, found wandering on other plants but with most individuals on a single host plant, and 'wandering weevils' probably simply moving between hosts;
3   others were known from few individuals and found on a variety of plants, so no specificity was detected;
4   other species occurred on a wide range of hosts, without undue abundance on any one of them, and they were presumed to be polyphagous generalists.

However, 48 species fell into the first two groups, indicating a preponderance of species likely to be at least moderately specialized. Thus, whereas conservation of some beetles is linked intimately with conservation of particular plants or other obligate food needs, this is not as critical for many others for which broader habitat features are at least as important.

Sites of special importance for conservation of beetles commonly transcend traditional biomes, but are often important for a combination of limited extent and particular categories of threat that influence either the resources available or the beetles themselves. In some cases, this separation is difficult to categorize simply. For the beetles of ERS (see p. 80) in southern Wales, a habitat of considerable importance in supporting numerous threatened species of ecologically specialized beetles, species richness was associated positively with stocking levels of cattle and sheep (Bates *et al.* 2007). Damage to the habitat by livestock is recognized widely in the UK, with riparian grazing having several effects, collectively influencing sediment debris, vegetation cover shading and aquatic food supply. However, Bates *et al.* also described direct trampling effects, following earlier appraisal by Sadler *et al.* (2004), whereby (i) sediment compaction could cause direct mortality and restrict availability of interstitial habitats; (ii) hibernation sites were destroyed; and (iii) defecation enhancing siltation of interstitial spaces and increasing the amount of nutrients there so that possible increased competition from additional (non-ERS specialists) might eventuate. The major outcome of this study was to recommend that stocking levels in areas occupied by high conservation value ERS beetle communities should be controlled carefully, and in some situations stock may need to be excluded entirely.

Habitats and resources in the landscape

The loss of habitats and resources, and changes in their accessibility and quality (reflected as carrying capacity for inhabitants), are of central importance in conservation and lead to a requirement to consider not only the features of these entities alone but also their dispersion and patterns, and hence their availability and accessibility, in the wider landscape. Much concern, for example, devolves on the consequences of habitat fragmentation to leave small patches in highly altered landscapes as perhaps the only places where formerly more widespread ecologically specialized organisms, including many beetles, can persist. The qualities of such remnants have received considerable attention, and the inferences vary widely according to individual species and contexts. The spatial dynamics of populations or species, and interpreting and understanding the patterns of these, pose considerable problems, as Talley (2007) indicated with her analysis of the Valley elderberry longhorn beetle (see p. 196) in relation to its host plant. More generally, the shape, size, structure and extent of isolation of habitat patches may all influence beetle richness and carrying capacity within that patch. Edges are transition zones between adjacent habitats and, as ecotones, constitute abrupt or more gradual changes between these. Often associated with fragmentation, edge effects at the junction of different natural or anthropogenic ecotones may be important influences on assemblage composition and dynamics. The edges themselves may support assemblages that include species not found in either of the main abutting habitats. All 19 forest fragments studied in Hungary by Lovei *et al.* (2006) supported edge specialist species of Carabidae, collectively comprising seven of the total 56 species taken across the edges, 41 of which were generalist (matrix) species, namely those found outside the fragments but not necessarily restricted to that region. In this survey, the edge was defined operationally as the outer 5 m of the patch, but in some other

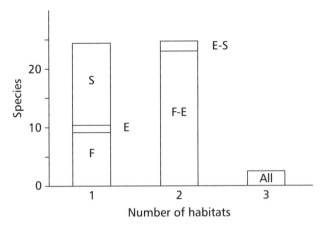

Fig. 3.2 Habitat specificity of dung beetles shown as the number of habitats in which species were collected over a forest–savannah ecotone. F, forest; E, edge; S, savannah. (From Spector & Ayzama 2003 with permission.)

studies it is not defined as objectively in spatial terms but, rather, operationally in terms of species' transitions. In beetle studies, agriculture–forest and natural savannah–forest edges have received more attention than others, with surveys comparing beetles from the two main habitats and the edge, and often demonstrating not only major turnover between the two systems but also peculiarities of the edge fauna. Thus for dung beetles examined in a Bolivian forest–savannah ecotone, almost complete turnover occurred between forest and savannah, with only two of the 50 most common species occurring in both (Spector & Ayzama 2003) (Fig. 3.2). The rank order of richness from the 73 species captured was forest > edge > savannah, with the species occurring on the edge were essentially a subset of the forest species.

Two further studies of similar distributional trends across ecotone borders are drawn from Finland (Heliola *et al.* 2001) and South Africa (Kotze & Samways 1999), both enumerating the numbers of carabid species across forest–edge–open ground transitions. In Finland, populations of forest-dwelling Carabidae did not decrease near the edge, so edges in themselves are not detrimental; inferences from South Africa were similar. The trends found by Heliola *et al.* may thus be more general in fragments, but do not address the effects of fragment isolation and the difficulties specialist beetles may experience in moving across intervening areas. Their main findings, reflecting relative incidence and abundance of 34 species, were (i) assemblages on the edge zones were more similar to those of forest interiors than to those in clearcut areas; (ii) there were no edge specialists, but interior forest species did not actively avoid the edge; (iii) despite richness and abundance being high in clearcut areas, those species only rarely penetrated to the forest interior; and (iv) some forest species were less abundant in clearcut areas. A representative distribution of an open-habitat species, showing attenuation towards the forest edge, is shown in Fig. 3.3; *Pterostichus adstrictus*, absent from forest, in total numbered seven on the edge and 110 in open clearcuts. Whereas Heliola *et al.*'s results were paralleled in Canada

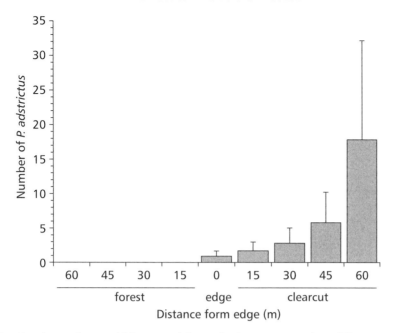

Fig. 3.3 Numbers of a carabid, *Pterostichus adstrictus*, captured at different distances from a forest edge into a clearcut area in Finland. (From Heliola *et al.* 2001 with permission.)

(Spence *et al.* 1996), so that northern boreal forests seemed to support no edge-specialist carabids, several such specialists were found in temperate forest in Hungary (Magura & Tothmeresz 1997) where, however, the edges were much 'softer', because of the presence of bushes, rather than being abrupt transitions.

Edge avoidance is demonstrated most easily by sampling relatively abundant species. Two species of bark beetles (Scolytidae: *Xylechinus pilosus*, *Cryphalus saltuarius*), assessed by presence/absence in 100-m transects into forest perpendicular to forest edges in Finland, clearly avoided the first 30 m of forest abutting the boundary. *Xylechinus pilosus* increased in frequency up to as far as 90 m from the stand border (Peltonen & Heliovaara 1998), while *C. saltuarius* was less heavily influenced by the edge. In a related study the attack densities of two other species of Scolytidae (*Hylurgops palliatus*, *H. glabratus*) increased substantially towards the forest interior, with bolts of *Picea abies* placed at different distances from the forest–clearcut edge used to detect the beetles (Peltonen & Heliovaara 1999). The trap log design (Fig. 3.4) yielded 10 bark beetle species, and results suggested that the effects of this edge on attack density and breeding rate differed significantly between species. The study sites did not include much natural breeding material, and one of the possible factors influencing attack rate is that ground contact creates different microclimates in moister interior forests than in sun-exposed logs nearer the edge. Some other bark beetles are known to avoid open forest edges, and exhibit preference for interiors that are commonly also darker and damper.

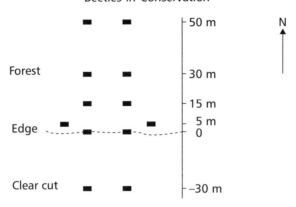

Fig. 3.4 Design for placement of sample wood (*Picea abies*) bolts (shown as black bars) across forest to clearcut edges to sample bark beetles across this ecotone in Finland. (After Peltonen & Heliovaara 1999 with permission.)

Such extensive edge effects may have serious implications for the size of effective core habitat for interior species in fragments. For example, Peltonen and Heliovaara (1999) suggested that it is likely that most specialized forest interior species may be no less susceptible to forest edges than these two common bark beetles. And for many small forest patches in Finland, the area of real interior habitat (namely that unaffected by changed edge conditions) is likely to be very small. The extent of interior habitat may be the critical regulator of both the total richness of specialist species and their abundance, but the extent of isolation is also very influential. For Carabidae in Ontario, less isolated forest fragments (those less than 2 km from others) had assemblages similar to those of continuous forest, irrespective of the fragment area (Burke & Goulet 1998). Nevertheless, within the overall 117 species of Coleoptera (of 13 families) in this study, the smallest fragments supported many fewer species (11) than the largest (55), with an average 52 species in continuous forest plots (Fig 3.5).

The above studies, and most other accounts of edge effects on beetles, imply short-distance effects across the edge, although most also have the important limitation that sampling has taken place over no more than a few hundred metres (at most) to each side of the edge. Should effects extend beyond this sampled range, increased concerns arise over the integrity, extent or even existence of interior habitat purportedly needed by specialist beetle species. This applies particularly in the smaller, remnant or fragment, areas commonly selected for invertebrate conservation purposes. Implications from a large-scale study of New Zealand beetles (Ewers & Didham 2008) pose serious doubts over how adequate such areas may be. Beetles were sampled along gradients extending 1024 m inside forest fragments and an equal distance into the surrounding grassland matrix in the Southern Alps in order to provide a sample of more than 26,000 beetles, including 769 species. Almost 90% of species responded to habitat edges (Fig. 3.6) and a number of them exhibited edge effects extending as far as 1 km inside forest patches, where beetle assemblages still differed in richness and composition from those in the deep forest interior. Large-scale effects may

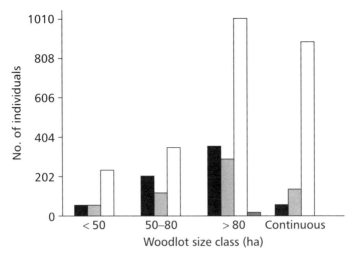

Fig. 3.5 Size classes of beetles collected from woodlots of different sizes in Canada. Body lengths of beetle size classes: small, 3–8 mm (black bars); medium, 8.1–15 mm (stippled bars); large, 15.1–30 mm (open bars); very large, > 30.1 mm (hatched bar). (After Burke & Goulet 1998 with permission.)

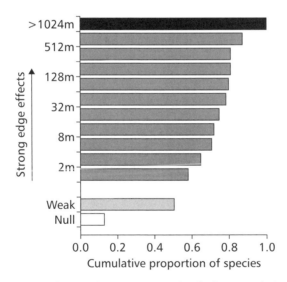

Fig. 3.6 Beetles across a forest edge in New Zealand: the cumulative proportions of the 78 common species responding at different spatial scales. Edge penetration distances represent distances over which habitat edges influence species abundance. Open bar, habitat generalists; light grey bar, species with significant but weak edge responses; dark grey bars, species with strong edge responses; black bar, responses for over 1 km). (From Ewers & Didham 2008, copyright National Academy of Sciences, USA.)

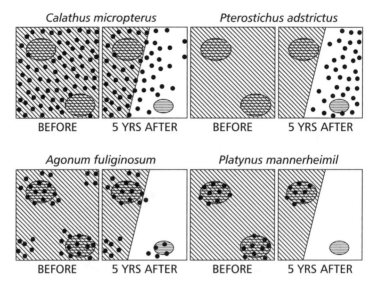

Fig. 3.7 Model showing responses to microclimate, edge distance and clearcutting within 1-ha blocks by four species of boreal forest Carabidae. Shaded sections are unlogged mature spruce forest (diagonal hatching) or spruce mires (horizontal hatching); unshaded areas are clearcut; black spots denote presence of species, with density of spots indicating abundance. (From Niemela *et al.* 1987, 1997 with permission.)

contribute to losses of species in supposed core areas and imply, as not widely acknowledged for insects, that specialist beetles and others may really need much larger areas than those commonly presumed to render them secure. Native forest strips 100 m wide do not necessarily support forest specialist beetles in Tasmania (Grove & Yaxley 2005) (see p. 186).

A consequence of habitat fragmentation is to increase the proportion of edge habitats whilst decreasing the proportion of interior habitat, so that the various edge effects due to climate alteration and changed species interactions are increased. Usually, the integrity of the interior habitat for beetles, and the patch size needed to assure habitat quality for them, are unknown. Edge effects extend variously towards the interior and in small patches (at least) may result in no part of the supposed interior habitat remaining free from their influence. Species responses, of course, differ. In boreal forests, some open-habitat carabids either do not enter forests at all or penetrate at most for only a few tens of metres (Niemela *et al.* 2007). Others, perhaps particularly some alien invasive species, penetrate deeply. The species most affected by fragmentation in this context tend to be forest specialists, because they may require very specific conditions that are degraded easily. Niemela *et al.* (1987) noted that the boreal *Platynus mannerheimii* in spruce mires requires wet sites within coniferous forests dominated by *Sphagnum* moss. Exposing mires by forest cutting led to disappearance of this species. The variety of responses of carabids to forest clearcutting is illustrated by the examples shown in Fig. 3.7.

Many of the surveys involving beetles compare the current fauna of remnant habitats with those of larger patches of that habitat (biome) in the same region, should these be available, and attribute differences in incidence and/or abundance to the process of habitat fragmentation, with the influence of edge effects, alien invaders, loss of interior habitat and isolation thwarting immigration commonly discussed. In many studies, sound pre-fragmentation data for the area are not available, so that the confounding effects of natural heterogeneity in beetle assemblages are not quantifiable. If a beetle species found in the parental assemblage is not found in the remnant, it is assumed to have been lost even though, at least in some cases, it may never have been there. A habitat patch, commonly a presumed or known fragment, comprises edge and core zones, and a general presumption is that ecologically specialized species typical of the parental habitat will continue to depend on largely unaltered regimes in the core while possibly coming into contact with other species to their detriment at and near an edge. Decreased patch size also relates to decreased heterogeneity of resources, so that any beetle species requiring a range of resources may be vulnerable to changes affecting any of these, and is likely to need a minimum habitat area for survival. As Magura *et al.* (2001) noted, a larger forest patch is likely to have greater heterogeneity than a smaller one and so support more specialist species. Among Carabidae, a number of individual species studies support this premise of diversity of resource needs, among them Magura *et al.* (2001) for *Carabus intricatus* (requiring dead and dying trees for overwintering sites) and De Vries and den Boer (1990) for *Agonum ericeti* (which needs conditions including cover of leaf litter and herbs, dead and dying trees, particular microclimates, and so on). Almost by definition, generalist beetle species are more versatile in their needs, but the above examples counsel against further fragmentation of already small fragments.

However, many studies indeed imply that impoverishment may occur within fragments. Lovei and Cartellieri (2000) investigated carabid assemblages in several patches of protected forest near Manawatu, New Zealand. A large forest tract yielded nine species in pitfall traps, whereas two putative fragments supported only two and three species, respectively, all of these present also in the larger stand. If the larger forest is the source of beetles for the fragments, lack of ability to disperse across the intervening regions might be associated with this impoverishment: seven of the nine species are wingless or short-winged and thereby poor dispersers. Predation by vertebrates might also be involved, with nocturnal mammals such as rats and possums implicated. Because these small forests were both natural and protected, Lovei and Cartellieri referred to these low-diversity arrays as 'collapsed assemblages'.

The latter study assumed that the vegetation profile of the remnants and parental site were similar, and contrasts with many in which vegetation differs clearly across treatments. A comparative study in New South Wales, Australia, likewise based sampling on stands dominated by white cypress pine (*Callitris glaucophylla*), using state forests and linear roadside remnants approximately 20 m wide (Major *et al.* 1999). Different assemblages occurred in the two treatments, but 64% (82 morphospecies in 29 families) of the forest species were found also in the linear remnants. This study implied that these remnant strips of woodland remain suitable habitat for many beetles of larger areas, and so are important both for connectivity and gene flow, and as reservoir habitats in the landscape.

The influences of forest fragmentation and related processes on beetles have only rarely been studied by purpose-designed long-term surveys. Two of these, undertaken in very different environments, are discussed below. The impacts on Carabidae in south-eastern New South Wales were appraised by pitfall surveys of replicated eucalypt forest plots of three sizes (0.25, 0.875 and 3.062 ha) within continuous eucalypt forest. Later, four of the six replicate triplets were retained as fragments in a newly established plantation of the alien softwood *Pinus radiata*, a tree that creates environments far different from those in the parental native forest. The remaining two triplets remained in unfelled forest as unfragmented control plots (Davies & Margules 1998). Pitfall traps yielded a pool of 45 carabid species, eight of which together comprised 92% of all individuals. The catches were accumulated at the 18 plots (totalling 144 sites), with an additional 44 sites within the pine plantation itself. These data allowed Davies and Margules to address several important general themes: (i) whether habitat fragmentation reduces carabid species richness; (ii) whether populations decline as a result of fragmentation; (iii) whether remaining subpopulations decline further on small fragments than on large fragments; and (iv) whether populations near the edges of fragments decline faster than populations in the interior. Habitat fragmentation did not alter species richness, but was associated with some changes in assemblage composition. Table 3.2 shows that three species were trapped only before fragmentation, and 12 (six of them only once and in the remnants) only after fragmentation. The most important outcome was the variation in response across the eight individual most abundant species. Three decreased in abundance in remnants compared with continuous forest, three others increased in abundance, and the other two did not respond to fragmentation. Of the six responders, only three responded further to fragment size: *Notonomus resplendens* was more abundant in small and large fragments than in medium-sized ones and was less abundant in all than in continuous forest; *N. variicollis* was more abundant in large and small remnants than in medium fragments or forest; and *Eurylachnus blagravei* was most abundant in large remnants and more or less equally represented in the other three treatments. The major finding was thus to confirm experimentally that different, even closely related or congeneric, species in an assemblage may differ individually and substantially in how they respond to a structural change. Each species may need individual investigation rather than relying on extrapolation from other taxa to clarify the nature of those responses. For forest carabids in France, Burel and Baudry (1990, see also Burel 1989) distinguished three groups of species according to their dispersal behaviour in relation to hedgerows extending from the forest core habitat.

1 Forest core species, found only close to the forest edge; their working definition for the boundary was within 100 m.
2 Forest peninsula species, using hedgerows to move as far as 500 m, but not found further than that distance from the forests.
3 Forest corridor species, using hedgerows as corridors and occurring at any distance from the forest, in this account as far as 15 km was noted.

Table 3.2   Habitat fragmentation effects on carabids: the Wog Wog experiment in New South Wales, Australia.

Treatment	No. of species trapped
Only before fragmentation	3
Only after fragmentation	12
Before in remnants and continuous forest	25
Before in remnants	20
Before in continuous forest	15
After in remnants and continuous forest	31
After in remnants	30
After in continuous forest	19
Pine plantation	31
Only caught in pines	9
Never caught in pines	5
Small remnants	23
Medium remnants	21
Large remnants	19
Remnant edges	28
Remnant interiors	24

Data shown are numbers of species in pitfall traps before and after forest fragmentation, in remnants and in continuous forest controls. Numbers 'after fragmentation' are given for the pine plantation matrix, for remnants of different sizes, and for remnant edges and interiors; note that there were twice as many trap sites in the remnants as in the continuous forest controls.
Source: Davies & Margules (1998) with permission.

A major study of effects of forest fragmentation on beetles in Amazonian forest investigated the extent to which beetles respond to habitat edges (caused by clearing of forest by selective logging or for cattle pasture) and patch size (forest remnant fragments of 1, 10 and 100 ha within a matrix of well-maintained pasture) (Didham *et al.* 1998a,b). Samples were based on Winkler extractions (over 3 days) of sieved litter from 1×1 m areas. Of nearly 1000 species captured, many were too scarce to evaluate, but the more abundant taxa encompassed a substantial array of different responses (Table 3.3). No consistent responses occurred within particular families or subfamilies, or within any given trophic group. Edge specialists and edge avoiders occurred and some responded to area of fragment. Likewise, some area responders were not sensitive to edges. Others responded to both parameters, still others to neither. Within each major category, trends in density tended to be positive (deep forest species) or negative (disturbed area species). For fragment size effects, relative species loss of the more common species was 49.8% (1 ha), 29.8% (10 ha) and 13.8% (100 ha), with declining density a precursor of these losses. One apparently contentious outcome from this study was the inference that common species were more likely to become locally extinct in small fragments than were rarer species, implying that rarer species (as better dispersers) may indeed be better at resisting disturbance from habitat destruction.

Table 3.3 Classification of responses of beetles to forest fragmentation.

---

1 Species responding to edge effects only (area insensitive)
  a Edge avoiders
  b Edge specialists
2 Species responding to area effects only (edge insensitive)
  a Large area specialists
  b Small area specialists
3 Species responding to both edge and area effects
  a Large area, edge avoiders
  b Small area, edge avoiders
  c Large area, edge specialists
  d Small area, edge specialists
4 Species showing no response to edge or area effects

Source: adapted from Didham *et al.* (1998a) with permission.

The Amazonian forest fragmentation study has become one of the more long-term studies of beetles and disturbance, with dung beetles a particularly informative focus. Klein's (1989) initial survey in 1986 (of three each of 1-ha fragments, 10-ha fragments and continuous forest plots) related smaller and fewer beetles (at lower population levels) in fragments than in forest to lower rate of dung decomposition in smaller fragments than in larger areas. Andresen (2003) resampled some of Klein's plots, and others, a decade later, when differences in species richness, general abundance and size of dung beetles among different size fragments were still evident. By 2000, Quintero and Roslin (2005) found that with regrowth of secondary vegetation between the original fragments, even to the extent of it forming closed-canopy forest at one site, those initial differences had largely disappeared. The secondary vegetation supported dung beetle assemblages similar to those in continuous forest, so that suitable beetle environments had been regained more rapidly than had been expected earlier. Part of the value of Neotropical dung beetles as indicators of forest change flows from the widespread supposition that most of the taxa evolved in forest environments. Since Halffter and Matthews (1966) noted that tree cover was the most influential factor on dung beetle assemblage composition, Halffter and Arellano (2002) endorsed that native forest species (in a study in tropical deciduous forest, rather than rain forest, in Mexico) undergo local extinction and are replaced by open-area species with changes to tree cover.

Habitat fragmentation effects and loss of carrying capacity are also reflected in the dependence of some beetles on particular successional stages, whether these are decay stages in rotting wood, the ages of vegetation stands or the changing variety of subclimax seral composition, any of which may be associated with very patchy distributions of specialized beetles. For example, the patchiness of the endangered *Carabus nitens* in fragments of the largest German heath landscape reserve (at Lüneberg) showed that this species prefers younger patches of *Calluna* heath over mature and degenerating heath. Despite the overall scarcity of the beetle, this apparent preference for low vegetation was considered to be an important reason for low beetle incidence (Assmann & Janssen 1999).

No *C. nitens* were found on heaths not managed to sustain pioneer/early-growth stages of *Calluna* but which instead were allowed to succeed to mature heath dominated by grasses. Such management to ensure the continual presence of early successional stages of *Calluna* was a critical aspect of maintaining habitat suitability for *C. nitens*. Fragment sizes less than 40 ha were considered probably unsuitable to sustain beetle populations, so that measures to increase the size of suitable patches and to increase connectivity were also recommended, a tactic possible only when such large areas as Lüneberg Heath are available.

Autecological studies are thereby needed to determine the mechanisms and processes involved in landscape-level habitat change on beetles, and the changes in resource supply often implicated in these. Habitat fragmentation may affect, for example, the supply of carrion or dung to burying beetles or dung beetles by changing the landscape suitability to the resident supplying vertebrates. Although small patches of habitat may continue to furnish all the needs of a beetle species, the smallest viable size is usually unknown. On the other hand, patches far larger than generally suspected may be necessary, for example some sensitive dung beetles may need forest patches of at least 85 ha (Larsen *et al.* 2008). Carrion (small vertebrate carcasses) is the critical resource for *Nicrophorus* beetles (Silphidae), four species of which are attracted to mouse carrion in parts of the USA (Trumbo & Bloch 2000), in areas where the beetles must compete with vertebrate scavengers and a variety of other insects for this material. As with the Australian carabids, different species of *Nicrophorus* apparently differ in their response to fragmentation in comparisons of numbers in woodlands and small and large fields, as measured by relative numbers attracted to bait carcasses. The beetles can disperse well, to judge from several studies on the endangered *N. americanus* (see p. 7), which has been documented as travelling up to 6.1 km in a night, although most (92% of 158 beetles) were recaptured within 1 km of their release point (Bedick *et al.* 1999). In Trumbo and Bloch's study, the greatest abundance and diversity of *Nicrophorus* occurred in woodlands, but *N. marginalis* occurred only in open fields (Fig. 3.8), from which two other species were absent. *Nicrophorus marginalis* did not come to carcasses placed at the edge of woods, and it appears reluctant to disperse across unfavourable woodland habitat. It was trapped only in large fields (> 25 ha) and never in small fields (< 5 ha), with this absence suggested to reflect that small patches cannot support a viable population. The other three species in this survey were considerably more abundant in woodland, but extended to the edge and one (*N. orbicollis*) also to open fields. A variety of factors may influence the distribution of the beetles, with greater exposure revealing the carcasses to other scavengers and perhaps reflecting rate of carcass decomposition. Increased woodland habitat (e.g. from reforestation) may thereby increase fragmentation of habitat for open-field/grassland species, and the converse. For *Nicrophorus*, habitat specificity reflects both macrovegetation features and their effects on resource supply.

Likewise, supply of the most suitable dung for specialized scarab dung beetles is sometimes critical. Forest dung beetles may not be able to extend their activity into human-made clearings for example, as the evolution of the dung-associated fauna may be almost entirely restricted to forests, as suggested by Howden and Nealis (1975, see also Peck & Forsyth 1982). Extensive clearing

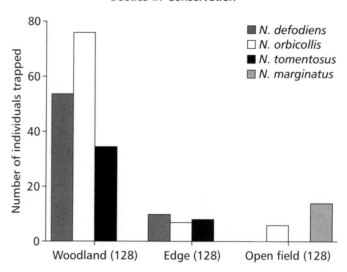

Fig. 3.8 Numbers of four different species of *Nicrophorus* captured at mouse carcass baits after 1 day in woodland, edge and open-field habitats in Michigan, USA. (From Trumbo & Bloch 2000 with permission.)

of forest may then be a potent threat to these beetles. The precise effects of such resource unpredictability and patchiness may extend to influences on complex behavioural interactions, as for Silphidae for which the unpredictable occurrence of carcasses in which they can breed suggests that they 'treat each reproductive event as if it were their only opportunity to breed' (Scott 1998, p. 596). Their communities can overlap considerably in habitat use, with soil composition and texture influential. Smaller species can dig more easily in damp organic-rich soil (such as in pine forests, as discussed by Gibbs & Stanton 2001), whilst larger species can operate within drier sandier soils (as in hardwood forests in the above study).

In an urban–rural gradient (below) of forest fragmentation in New York, species richness of burying beetles and beetle abundance were reduced by one-third and two-thirds, respectively, in fragmented forests (Gibbs & Stanton 2001). Ten species were captured, with sites in contiguous forest typically yielding six or seven species and those in fragmented forest harbouring only three or four species, with these tending to be the smaller-bodied habitat generalists. As in some other studies of beetle assemblages, compositional changes tended to reflect changes in a few predominant species. In this instance, *Nicrophorus defodiens* decreased by around 10-fold from contiguous to fragmented forests, whereas *N. tomentosus* more than doubled in abundance from contiguous to fragmented sites. Part of these changes may be attributed to differences in carcass supply, these also relating to several factors such as the source of supply, and the exposure of carcasses to scavengers such as crows and opossums as possible competitors for the resource. In some carrion beetles, exposure of the carcasses may simply link with temperature differences that mediate the outcomes of competition between *Nicrophorus* species (Wilson *et al.* 1984).

In many cases, the intervening altered areas in a landscape are viewed as barriers (see p. 75), with this between-patch matrix deemed unsuitable for many species to inhabit and restricting or stopping movement between the suitable patches. A ski trail (19–22 m wide) established in Vermont, USA in 1968 is mown infrequently (less than once per year) to maintain it as a permanent opening free of tree encroachment from the adjacent balsam fir (*Abies balsamica*)-dominated forest. Studies of Carabidae (37 species) and Elateridae (20 species) trapped on the ski run itself, in the parental forest and along the edge between these showed substantial differences (Strong *et al.* 2002), with similar patterns in the two families. Each of the three zones supported distinctive assemblages, and most species in the ski trail itself were colonizers from lower altitudes, presumably using the cleared area as a corridor along which to expand their range upwards. Only 2 of 540 individuals of seven forest species occurred in the open areas, and only 2 of 173 individuals of the five open-habitat species were trapped in the forest. In this example, the edge constituted an important separator of these assemblages and despite richness and diversity of these beetles being rather similar each side, turnover within the assemblages was high. From this, Strong *et al.* (2002) suggested that uncritical fragmentation of this sort, particularly by features running perpendicular to the climatic gradient, may place isolated beetle populations under increased risk.

Some fragmentation effects have been evaluated for European carabids from studies assessing beetle richness and abundance in forest remnants and forest islets separated from the larger remnants by a road highway (Fig. 3.9), with samples taken also along the road verges abutting these (Koivula & Vermuelen 2005). This study, also incorporating mark–release–recapture surveys (see p. 62), inferred that (i) smaller patch size (islet area) negatively affects forest carabid catch and the overall species richness; (ii) habitat type can affect the intensity of this patch size effect; (iii) carabid assemblages of the forest sites vary with traffic volume; (iv) forest carabids only rarely cross highways (see p. 75); and (v) open habitats, here road margins, are barriers to the dispersal of forest carabids.

The wide array of European carabids appraised by Rainio and Niemela (2003) led to partial clarification of some general impacts of fragmentation, with some species resultantly threatened. The most vulnerable species were strict forest specialists with limited capability for movement (mostly through being flightless) and species found in the interior of fragments. The general impacts were listed by Rainio and Niemela as (i) species composition changing even though species number might not; (ii) species abundance changes (upwards or downwards) but not in all species; and (iii) specialist species decline whilst openhabitat species increase. The implications of movement modes were revealed from the presence of brachypterous open-ground carabids on small isolated patches of farmland in boreal forest of Finland (Kinnunen *et al.* 1996, 2001). The occurrence of *Carabus cancellatus* (on three patches) and *Dyschirius globosus* (on five patches) was unexpected and three possible explanations for their presence were advanced: (i) movement along road verges as corridors; (ii) transport with people, in vehicles or with hay or earth; and (iii) as small relictual populations surviving from times when the landscape was historically

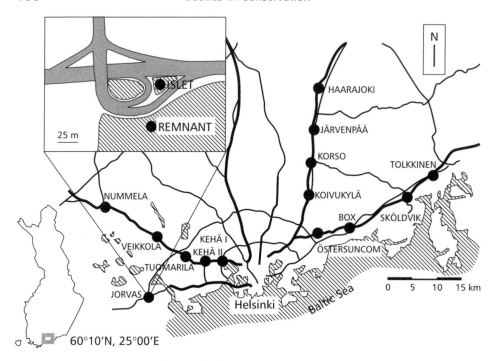

Fig. 3.9 Sampling design for a pitfall trapping exercise for Carabidae in Finland, illustrating the difference between islet and remnant habitats in relation to road design. (From Koivula & Vermuelen 2005 with permission.)

more open. Road transport was suggested also as a mode of arrival for generalist open-habitat carabids in small cleared areas only a few months after forest logging (Koivula & Niemela 2003).

Small relatively natural fragments may be important refugia, even when isolated from larger patches of parental habitat. In New Zealand, Watts and Lariviere (2004) showed that small urban reserves in Auckland supported fewer beetle species than larger forest patches, but were still important reservoirs of beetle diversity in heavily altered landscapes. Numerous beetles of individual conservation interest are known only from small fragments within a suspected original larger range, with their incidence across a number of separated patches taken as suggestive of formerly being present in the now alienated intervening area. As one such example, the elephant dung beetle *Circellium bacchus* is now 'restricted to tiny fragments of its putative original range' and has undergone 'a massive decline in the species' range' (Kryger *et al.* 2006), essentially to within the xeric southern and eastern Cape, with its flightlessness probably furthering its susceptibility (Chown *et al.* 1995). Early records of this notable species indeed demonstrate a formerly wider southern African distribution, but nevertheless still one where small localized populations occurred rather than being found continuously across the landscape. In reserves (Addo Elephant National Park, Buffalo Valley Game Farm, and four smaller regions along the southern Cape coast), Kryger *et al.* recorded *C. bacchus* as 'restricted to areas of dense undivided vegetation' that also harbour large herbivores that provide the dung used by the beetle. Equally, for many beetles, conservation of the remaining inhabited fragments presumed to be true remnants is a keystone of any management plan, with the single to few such populations commonly needing site-specific attention, in addition to any overarching management plan.

Mosaics of land-use patterns are clearly valuable in fostering beetle diversity, with each stage in gradients of vegetation type or disturbance (e.g. from conversion of natural forests to plantation or pasture) generally yielding species not found in any other stage sampled. This general pattern has been found repeatedly in numerous studies in many parts of the world. In a Costa Rican study by Gormley *et al.* (2007) based on pitfall trapping of 422 beetle species in 26 families, assemblages in more highly disturbed sites were more similar to each other than to those in logged or primary forest. Any level of disturbance signalled decline in richness of forest species, so that the mosaic needed for effective conservation must include undisturbed and minimally disturbed forest patches.

Studies of the kinds noted above have been important in helping to understand the ecology of fragmentation effects. A complementary approach is to examine genetic differences between populations on different habitat remnant patches and to attempt reconstruction of their historical biogeography, incorporating any knowledge of landscape changes. This approach also draws on knowledge of dispersal capability as reflecting the extent of population separations. The brachypterous forest ground beetle *Carabus auronitens* can use hedges as corridors for dispersal, and its genetic pattern (Drees *et al.* 2008) helps to demonstrate impacts of human activity on the landscape in the area of Germany surveyed. Both fragmentation and defragmentation effects were shown, the latter representing processes of increasing connectivity between populations and so helping to reduce genetic differentiation between populations by providing for more frequent contact. Drees *et al.* postulated a history of multiple isolated refuges and colonization/recolonization events in this species, with defragmentation achieved largely through a linear network of hedgerow corridors. Their historical reconstruction of landscape change has implications for many species. The large suitable woodland habitat areas present formerly were changed through clearing, some to heathland or brushwood areas, but leaving no more than about 10% of the area forested, so that conditions for species such as *C. auronitens* declined substantially. Substantial reforestation occurred from the early 19th century, with directives to landowners to plant hedges, leading to establishment of a dense network by the early 20th century. These enabled *C. auronitens* to move from its refuges and colonize the new forests. Drees *et al.* (2008) commented that this occurred around the beginning of the 20th century, with the consequent 'secondary contact' a major factor in explaining the genetic patterns they found across 26 populations.

Habitat gradients for beetles

The gradual transitions between biomes across landscapes create gradients of species incidence and assemblage composition in response to vegetational, climatic, altitudinal and other changes that may be viewed as forming a continuum between extremes, but with the gradual transitions sometimes rendered abrupt by other changes in the landscape such as those involving creation of 'hard edges' (see p. 90). Grassland–forest, lowland–highland or wetland–dryland transitions all fall into gradients that can be relatively natural, with each differing site along them supporting species and assemblages that are in some way specific to the particular set of conditions present.

Changes by people create a variety of, sometimes more abrupt, other transitions, so that gradients then extend between more natural and more anthropogenic conditions. Beetle distributions along any such gradients may reveal features that restrict hospitality, the extent of dispersal and the levels of ecological specialization and associations present, and thus provide information of considerable value in conservation. Carabidae in the northern temperate regions have been the focus of many such surveys at points along various environmental gradients, and across ecotones of varying levels of difference. Gradients of soil humidity, for example, correlated with differences in assemblage composition of Carabidae in Norway (Ottesen 1996).

Urban–rural gradients have been of particular interest in recent years (Niemela *et al.* 2000) in helping to incorporate urban ecology with more natural regions, and prompting development of a broad standard sampling regime for carabids projected to allow comparisons along such gradients in different parts of the world. The Globenet project has since led to several analyses of the carabid fauna and its transitions along urban–suburban–rural gradients, in widely separated cities, and has helped to elucidate any general patterns of the effects of urbanization on Carabidae. Much, of course, depends on features of the local urban environment. One purported hypothesis/generalization is that carabid abundance and species richness are lower in urban areas than in more natural areas, and this appears to hold for several studies: Helsinki (Finland), Edmonton (Canada) (Niemela *et al.* 2002) and Hiroshima (Japan) (Ishitani *et al.* 2003), the exception being Sofia (Bulgaria) (Niemela *et al.* 2002). As Ishitani *et al.* noted, gradients such as these are unlikely to be simple or uniform, with the beetles' responses influenced by many different factors reflecting the local environment. In an urban–rural transect in Helsinki (Alaruikka *et al.* 2002), the intermediate suburban sites were richer in Carabidae (24 species) than either urban (18 species) or rural (17 species) sites. However, site differences were substantial, and the gradient categories were not significantly different in either beetle richness or abundance. This study suggested that urbanization along the 20-km sampling distance had few major effects on beetle assemblages, although some individual species changed in different ways according to body size (larger beetles less likely to occur in urban forest) and ecological specializations. For Carabidae, large flightless woodland specialists may be especially susceptible to changes from urbanization (Sadler *et al.* 2006).

The interplay of ecological specialization, body size and dispersal capability along gradients can become complex, and another possible generalization from Globenet studies is that disturbed areas are characterized largely by small-sized and generalist carabid species, linked possibly with small ground beetles often being more vagile than larger ones, because they include higher proportions of winged species. On that premise, the chances of collecting smaller carabids should increase in urban areas, as their frequency increases. The third proposed generalization was that forest specialists (rural areas) should decrease towards cities, with urban forests (remnant patches) supporting fewer such specialists than natural forests.

The Edmonton survey (Hartley *et al.* 2007) emphasized that urban areas are commonly highly disturbed and support early succession vegetation, again demonstrating their greater suitability to generalist carabids. Findings of that survey included (i) that native carabids were less abundant in urban than in suburban and rural sites; (ii) brachypterous carabids were proportionally more abundant in the urban regions (33% of the total) than suburban (10%) or rural (5%); and (iii) marginally significant responses to the different zones occurred in wing-dimorphic and macropterous taxa. The two carabid species found only in the urban area were both represented by singletons, and the three species found only in rural area were also scarce (with 1, 4 and 5 individuals, respectively). All the more abundant species (10 or more individuals) occurred in at least two treatments. At each stage of a habitat gradient of this kind, intensity of management may be influential. In Canada, carabid assemblages were compared along an urban–rural gradient, as above, on unmanaged grassland (unmowed, dense tall grass vegetation) and highly managed grassland (regularly mown graveyards). Graveyard assemblages had fewer species (Hartley *et al.* 2007) and catch rates, but three introduced species (*Pterostichus melanarius*, *Carabus granulatus*, *C. nemoralis*) were by far predominant in both treatments, so that the pool of 21 native species comprised only 14% (355 individuals of 20 species) and 21% (136 individuals of 12 species) respectively of catches on the unmanaged and managed sites. Brachypterous carabids were proportionately less abundant in graveyards. The differences possibly reflected higher habitat complexity in the unmanaged grasslands. The three rural-only species were all scarce, two of them represented by singletons as likewise were the two urban-only species. Native carabids were considerably scarcer in inner urban areas, in marked contrast to abundance of the two alien species of *Carabus*, both of which were most abundant on the inner urban sites. The four major effects of urbanization implied from this survey by Hartley *et al.* (2007) were (i) the preponderance of some introduced species on urban sites; (ii) the lesser abundance of native carabids on urban sites; (iii) increasing road density negatively affected native and alary-dimorphic carabids in graveyards, but positively affected brachypterous carabids in graveyards; and (iv) the more heavily disturbed graveyards supported poorer assemblages than other sites. The first two of these are relevant to the largest scales along the gradient, point (iii) relates to a scale of up to a few square kilometres and point (iv) relates to within-patch (habitat quality) effects. One practical outcome is confirmation that unmanaged grasslands indeed support more carabids than highly managed ones, so that ecologically oriented management of, for example, roadside grassy verges (Eversham & Telfer 1994) is preferable to simple repeated close mowing for aesthetic reasons.

Other gradients of land use, for example involving the extent and intensity of agriculture and forestry, have also attracted considerable attention in relation to beetles, and carabid assemblages can change along these in relation to

Table 3.4 Ranked abundance (percentage of total in parentheses) of the five most abundant species of Carabidae captured in two consecutive years across a gradient of land-use intensity in Scotland.

Species	Number	
	2001	*2002*
Nebria brevicollis	3059 (17%)	5531 (26%)
Pterostichus madidus	2993 (17%)	3192 (15%)
Agonum muelleri		1537 (7%)
Bembidion tetracolum	1334 (8%)	1168 (6%)
Anchomenus dorsalis	1164 (7%)	
Calathus micropterus	1030 (6%)	
Pterostichus niger		964 (5%)
Total proportion	55%	59%

Source: data from Vanbergen *et al.* (2005).

soil and vegetation properties in particular. This category is one of the major anthropogenic transitions in terrestrial environments, and most commonly has involved the change from forest to cleared agricultural areas, whether pastoral or cropping, or to silvicultural activities, both commonly also involving alien species (see p. 124) in changes widespread across landscapes. Heterogeneity of disturbance/change or of the varying components of landscape structure (such as vegetation) may have substantial effects, with information on these valuable in planning conservation management.

A gradient of six 1 × 1 km quadrats in Scotland, with increasing proportions of agricultural land, were pitfall sampled for carabids in two successive years (Vanbergen *et al.* 2005), yielding substantial numbers (17,494 individuals in 51 species in 2001; 20,935 individuals in 54 species in 2002) (Table 3.4). The five most abundant beetles in each year comprised 55% and 59% of all individuals. The most abundant species in both years, *Nebria brevicollis*, showed a clear preference trend to the arable end of the gradient. Although this species is commonly thought of as eurytopic, the numbers in each quadrat along the arable–forest gradient (i.e. 3900, 3287, 1211, 173, 18, 1) suggest otherwise. *Pterostichus madidus*, the second ranking species in both years, might have responded to some level of intermediate disturbance, with its numbers (465, 1054, 416, 4103, 147, 0) peaking in a forest-dominated mosaic quadrat. In both years, the apparent trend was of increased species richness with decline of forest towards open agricultural areas, with the highest richness in either the arable site (2001) or the grassland-dominated site (2002), and the semi-natural forest the poorest in both years.

Drawing from cases such as those cited above, agricultural intensification has been considered a key threat to many invertebrates, through changes to natural vegetation and impacts of agricultural chemicals. Many beetles along land-use gradients are demonstrably somewhat resistant to disturbance, so that many taxa not dependent on particular plant species may persist in disturbed agricultural landscapes. However, more general concerns over effects of intensive agriculture on natural biodiversity have stimulated many surveys of such insects and of the factors that harm or apparently benefit them (Verdu *et al.* 2007). In general, those surveys comprise comparative sampling of particular focal beetle groups in sites with differing intensities and modes of farming practice. For dung beetles in Ireland, the faunas of intensive farms (good-quality soils, high inputs of synthetic fertilizers and pesticides, anthelmintic treatments of cattle), organic farms (good-quality soils, no input of synthetic chemicals, no anthelmintic treatments) and rough-grazing farms (poor-quality soils, low input of synthetic fertilizers and pesticides, some anthelmintic treatments of cattle) showed some benefits of organic farming for the beetles (Hutton & Giller 2003). Organic farms had significantly greater beetle richness, diversity and biomass than the other two treatments.

A rather different perspective on gradients emerges from several studies on altitudinal differences in species incidence and assemblage structure in Scarabaeoidea. As for carabids, any ecological generalizations may be thwarted by individual site characteristics, but general points of conservation value emerge. For example, three separate altitudinal transects in the Neotropics all showed decrease in scarab species richness with increased altitude (Escobar *et al.* 2007). In each of these, forested and cleared cattle-grazing sites were sampled at each altitude in order to compare assemblage constitution in parallel in these two systems. The three localities differed considerably in details of species turnover and the constitution of the local species pools. The changes resulting from transforming forest to pasture also differed clearly at each.

The studies cited above deal with particular families of beetles, and surveys of wider arrays of beetles are relatively scarce as well as being correspondingly more complex and ambiguous to interpret. Beetles sampled by pitfall and window traps along a forest productivity gradient (comprising sites in spruce mires and herb-rich, mesic and subxeric forests in Finland) included 435 species across 52 families, with 98 species represented by singletons (Simila *et al.* 2002). Catches were dominated by one species (*Zyras humeralis*, Staphylinidae), which comprised 70% of the 100,333 individual beetles captured.

Gradients of anthropogenic change (emphasizing 'space') and gradients of natural succession (implicitly incorporating 'time') have many parallels, so that patterns along each may have features in common, and individual variety. The patterns of distribution of saproxylic beetles along the decay gradient on different trees is one studied example. In Quebec, Canada, 80 snags each of black spruce (*Picea mariana*) and aspen (*Populus tremuloides*), the dominant trees in northern boreal forests in the region, were examined for the beetles present (Saint-German *et al.* 2007). One metre lengths of the snags, representing four stages of decay (categorized by measurements of wood density) were dissected by hand and the incidence of Buprestidae, Cerambycidae and Scolytinae assessed. Of 17 beetle taxa in spruce, 15 were restricted to early decay stages. In contrast, few species occurred in younger snags of aspen, with richness and

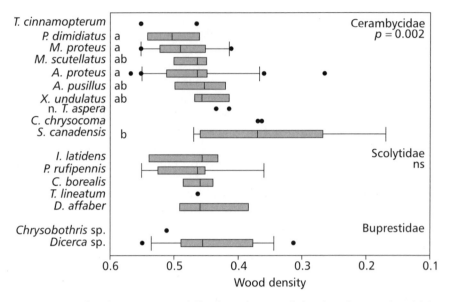

Fig. 3.10 Example of resource specialization: the wood density of snags in which various saproxylic beetles were found on black spruce in Canada. Letters indicate significant differences between mean wood density of hosts of different species. (After Saint-German *et al.* 2007 with permission.)

abundance increasing at later decay stages. The patterns of stressed host species and dead host species differed across these two important host species. The combined patterns for Cerambycidae are summarized in Fig. 3.10.

Assemblage composition differences along gradients, as well as including ecologically gradient taxa, can also encompass interdigitation of different biogeographical elements, although this theme has not been emphasized strongly in beetle studies. An altitudinal survey of dung beetles (with only 2 days of sampling yielding a pool of 74 species for evaluation) over about 2350 m in South Africa revealed three discrete communities with clear altitudinal zonation (Davis *et al.* 1999) and based on different biogeographical elements. Two of the communities (coastal escarpment forest at 500 m, at the lower end of the gradient, and coastal to highland gradient spanning 500–1500 m) each comprise high proportions of east-coast endemic beetles (abundance > 89% in coastal forest) and montane endemics (abundance > 84% in montane grassland). The third community in these zones is more heterogeneous, with predominant taxa being high-veld endemics and others with distributions extending to the tropics. Highland endemics peaked at higher altitudes, and the more tropical elements were proportionately more abundant at lower levels. Table 3.5 shows that abundance was low in lowland forests but increased to its highest level at around 1500 m, where species richness was also greatest, above which both parameters declined. No individual species occurred at both extremes, so that in this study altitudinal changes involved substantial turnover and differences strongly reflected biogeographical affinity and pattern.

Table 3.5 Abundance, species richness and diversity of dung beetles across a vegetational/altitudinal gradient in South Africa (southern KwaZulu Natal).

	Forest 500 m	Grassland					
		500 m	1000 m	1500 m	1900 m	2400 m	2800 m
Species richness	13	33	23	38	22	11	13
Mean species	3.7	13.2	5.5	20.4	11.3	2.8	3.1
per trap (SD)	(1.7)	(2.7)	(4.4)	(2.5)	(1.5)	(1.6)	(1.5)
Mean abundance	16.6	38.9	31.2	369.7	158.6	6.5	12.5
per trap (SD)	(15.6)	(17.1)	(27.7)	(134)	(45.5)	(5.1)	(11)

SD, standard deviation.
Source: after Davis *et al.* (1999) with permission.

Remnant habitat values: brownfield sites

Urban derelict areas, commonly small and consisting of only a few hectares, are sites changed by human activities and later abandoned to constitute open spaces in predominantly urban developed regions. They include sites on which buildings have been demolished, industrial dumps, landfills, sand or gravel pits, mine sites and other wasteland (Small *et al.* 2003; see also Eversham *et al.* 1996) (see p. 162), for which the general epithet 'brownfield sites' has been coined. In Britain, brownfield sites are important habitats for beetles, and support populations of some rare and scarce species (Eyre *et al.* 2003b). They are regarded as refugia for many species of conservation concern. Study of 78 sites in England suggested that vegetation cover and site drainage were major determinants of the distribution of Carabidae and Staphylinidae. For several decades after the Second World War, bombsites in London and other European cities were sometimes regarded as important local sanctuaries as successions developed and enabled progressive colonization by numerous local animals and plants. Even very small sites can accumulate substantial richness (as found in a number of studies of garden fauna by Welch 1990). More recent studies, some of them on Carabidae, have emphasized the values of post-industrial sites for insect conservation. In a survey of 26 brownfield sites in Birmingham, UK, 63 species of Carabidae included two nationally scarce species (Small *et al.* 2003). Some of the trends in assemblage composition are summarized in Fig. 3.11, reflected under the four main successional changes delineated. Carabid assemblages on derelict sites are related to within-site variables such as site age, substrate and vegetation, rather than landscape geometry, although isolation of some sites may affect rates of colonization by less dispersive species, so that more proximal sites may be important in sustaining populations of rare species. The major management recommendation by Small *et al.* (2006) was to focus predominantly on habitat quality, such as by maintaining early successional stages with a diversity of annual plants and enhancing substrate variety.

Reclamation of industrial areas, such as those used for open-cast lignite mining in Germany (Brandle *et al.* 2000), and abandonment of such areas are

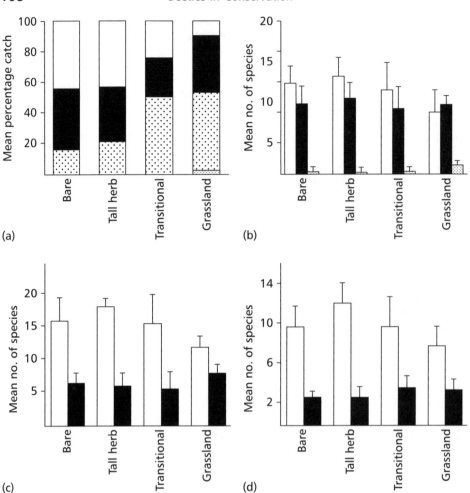

Fig. 3.11 Variations in species characteristics in the assemblages of Carabidae along a succession gradient from bare ground to grassland on urban derelict sites in England, including a predominant species (*Pterostichus madidus*) amongst three other main ecological groupings. (a) Habitat preferences including *P. madidus*: open bars, open-country species; black bars, generalist species; stippled bars, *P. madidus*; shaded bar, wood and damp species. (b) Habitat preferences excluding *P. madidus*: open bars, open-country species; black bars, generalist species; stippled bars, wood and damp species. (c) Flight ability, excluding *P. madidus*: open bars, capable of flight; black bars, flightless. (d) Proportion of phytophages (open bars) to carnivores (black bars), excluding *P. madidus*. (After Small *et al.* 2003 with permission.)

both options used in practical land management. Effective monitoring is a prerequisite for assessing the outcomes of restoration in order to interpret its success, but assessing the values and trajectories of abandoned areas is also of considerable interest. Carabidae have become valuable indicators in such contexts. A pitfall trap survey of two non-reclaimed lignite mines and a dump in

Saxony yielded 203 beetle species, representing 27 families, across 21 sites. All sites had been abandoned for several decades, and when sampled represented a variety of successional stages from dry bare ground to grassland and early birch woodland (Brandle *et al*. 2000). Carabidae (with 75 species) was the richest family found. It included about 25% of the 4099 beetles taken over five summer months of trapping. Their richness was higher than in nearby agricultural areas, probably reflecting the higher habitat diversity across the mining sites. Ten of these Carabidae were among those listed as endangered for Saxony and, as for many other studies on beetles, the two major elements comprised generalist species extending to these sites as part of their normal range, and more stenotopic species suggested to use these sites as surrogates for more specialized habitats lost from the area at an earlier time. The abandoned mine sites were regarded as having high conservation value for Carabidae in areas that were otherwise used for extensive agriculture. Conservation of such sites with early successional stages was advocated, with the comment that restoration by afforestation and flooding of mining areas would lead to reduced habitat diversity.

Set-aside and similar schemes, whereby formerly productive agricultural land (sometime used for intensive crop production and subjected to heavy fertilizer and pesticide applications) is removed from production and becomes (at least temporarily) part of the conservation estate, in some ways parallel the context of brownfield sites. Monitoring set-aside areas has demonstrated their significant conservation values, with several studies on heathlands and heathy grasslands in Europe revealing rich carabid assemblages and the presence of individual species of conservation concern. In eastern England, Telfer and Eversham (1996) noted that set-aside arable sites had potential to conserve much of the carabid fauna of natural biotopes in that region. Year-long surveys of three set-aside fields in Belgium yielded 11 red data book species for Flanders (Desender & Bosmans 1998), including one (*Amara tricuspidata*) rated as critically endangered and two (*Harpalus froelichi* and *H. griseus*) as endangered, implying that these sites can accrue considerable conservation significance. The role of set-aside areas, and the related theme of habitat offsets, need much further investigation as tactics in beetle conservation.

Islands and island habitats

Islands

Islands are habitats of particular interest to conservationists, biogeographers and evolutionary biologists alike, and are paralleled functionally by the many continental habitats regarded as isolated fragments within a landscape, with the true islands of this chapter separated by water (rather than by inhospitable terrestrial matrix) from other land. Most studies on island beetles have been made on conventional islands that have existed for a long time. Occasionally, however, opportunities arise to gain additional ecological insight by studying beetles on islands of much more recent origin. One notable case is the Lago Guri Hydroelectric impoundment in Venezuela, where a dam built in 1986 flooded

4300 km^2 to leave an archipelago of isolated forest islands protruding from the lake. Dung beetles sampled over four years on 33 of these islands ranging in area and extent of isolation (distance to shore) showed that species richness, density and biomass all decreased sharply with increasing isolation and decreasing island area (Larsen *et al.* 2008). Ecological variety was reflected in designation of four response groups among the 32 species collected. The islands were categorized into five classes on species richness, and each beetle species then allocated to a response group based on its mean relative density plotted against island richness class. Forest beetles comprised three groups.

1   Sensitive: species declining in density from the most species-rich class and found only in one or two island classes.
2   Compensatory: species that increased in density in at least one lower richness class.
3   Persistent: species declining in density as in sensitive species, but occurring in three or more richness classes.

The fourth category included those adapted to open areas and so likely to differ in their responses to forest fragmentation from forest species. They were designated 'supertramps' to reflect their habitat breadth (Larsen *et al.* 2008).

Older islands have been the foundation for many studies of colonization, evolution and speciation (commonly leading to very high levels of endemism on isolated oceanic islands over prolonged periods of isolation) and, in turn, the threats to these biota resulting from invasive species and wider despoliation of island environments. Any more general threat may become relevant, but the small size of many islands may render some stochastic events (fire, volcanic activity, flooding, landslides) more destructive than for similar-scale events on larger continental areas. Such studies involving island beetles have contributed considerably to ecological understanding. Threat intensity on small islands can relate strongly to scale. Initial areas of occupancy may be extremely small and extirpations or extinctions consequently more readily achieved by disturbances that would be absorbed or buffered across larger or less isolated arenas. This alone may make a compelling case for conservation of many island beetles, but the general points made to merit the 47 species of *Rhyncogonus* on Hawaii for conservation (Samuelson 2003, p. 5) could apply to many other taxa.

1   *Rhyncogonus* exemplifies an island radiation in a hotspot area containing many endemic species in great danger of extinction.
2   The genus ranges throughout the archipelago, with most of the flightless species endemic to single islands.
3   The genus contains many apparently rare species and one believed to be extinct.
4   Many of the species are provincial, i.e. restricted to small areas on an island.
5   Lowland and montane species are both represented.
6   The genus is a dominant (here, leaf-chewing) group on the archipelago.
7   *Rhyncogonus* has been used in evolutionary and dispersal studies and interpretations.
8   *Rhyncogonus* is generally aesthetically attractive.

Remote oceanic islands and archipelagos usually have faunas that are unbalanced (disharmonic) in relation to those of mainland source areas. As Peck (2006) noted for the Galapagos Coleoptera, only a subset of mainland families and a very limited subset of subfamilies and tribes found on Neotropical mainland areas are present on the archipelago. Reasons for this widespread pattern include likelihood of long-distance dispersal and opportunities for this to occur, and the presence of suitable resources and receiving environments on the islands. Peck characterized the beetles present on the Galapagos as 'able dispersalists, rugged colonists, and adaptable in acceptance of available microhabitats and food materials'. Whereas continental islands, by definition at one time part of continental landmasses, may retain components of the original continental fauna, so that the chance presence of poorly dispersing or otherwise specialized species may ensure their initial presence in the island community, all species on oceanic islands must arrive. Peck's (2006) comment above applies to faunas that have evolved independently of any previous continental physical connection.

Considerable differences in faunal richness may occur between tropical islands and those at high latitudes, and the latter may at times be particularly useful in studies of ecological and evolutionary patterns simply because of this simplicity whilst still being very isolated environments (Chown 1994). Chown's focal example was the group of flightless weevils comprising *Ectemnophorus* and its allies, a suite of only 36 species but one of the more diverse faunal groups on the Province Islands in the southern Indian Ocean. This monophyletic group comprises around 80% of the Coleoptera on these remote islands, and has proved a useful tool for studying historical biogeography. The weevils are related most closely to some within the New Zealand region. Six genera are involved, and many of the beetles are entirely cryptogam-feeders whilst others eat angiosperms; most species are polyphagous. However, habitat use corresponds closely with diet, so that many species are restricted to quite specialized epilithic biotopes (Chown 1992, 1994). Intriguingly, this may leave angiosperm-frequented areas of the islands relatively beetle-free (Kuschel & Chown 1995).

Different dispersal modes for oceanic island beetles may be widely differentially available to the various potential colonists, and impose a substantial taxonomic and ecological filter on the pool of candidates. The four main categories of dispersal, as delineated by Peck, are (i) aerial transport, by flight or passively by wind; (ii) 'rafting' on floating materials or swimming directly (see also Bell 1979 on Rhysodini carabids); (iii) hitchhiking on or in other organisms; and (iv) accidental or deliberate introductions by people. For the Galapagos beetles, the first two categories have been predominant, with sea surface transport available to flightless beetles and larvae as well as winged adults, and aerial transport largely for winged adults. Transport on animals is restricted to a few parasitic taxa, while human introductions are considered (as elsewhere) likely to increase to become the most serious threat in conservation of remote island biotas, notwithstanding continued and accelerating quarantine endeavour. For the flightless *Rhyncogonus* weevils on Hawaii (see above), Samuelson (2003) commented that 'Seabirds have undoubtedly had a role in . . . dispersal', based in part on parallel distributions of some *Rhyncogonus* species and certain migratory birds, particularly ground- or burrow-nesting petrels and shearwaters.

Island endemic beetles are, of course, by no means immune from further dispersal at any time after their initial arrival, so that patterns of distribution and evolution within archipelagos have attracted considerable interest. Interpreting causes of island extinctions is simply one facet of this. Across islands in the Mediterranean area, Tenebrionidae are regarded as an important indicator of land use, with the benefit that they can be surveyed easily by pitfall trapping and direct hand-collecting. Endemism is not uncommon, but island extirpations from human activity are known on several islands. For example, losses on islands to the east of Spain include tenebrionids, with the causes attributed as urban activities, introductions of alien species, and tourism (Cartagena & Galante 2002). The alien predators, as elsewhere, include rats and poultry. The flightless Darwin's darkling beetle (*Ammophorus insularis*, Tenebrionidae) was carried from the Galapagos to the Hawaiian Islands in the early 19th century and has since then been transported to all major islands in that group (Peck 1997).

Flightlessness has developed many times among island insects and whereas winged beetles commonly occur on more than one island in an archipelago, greater levels of speciation tend to be found in the less vagile groups. Flightless single-island endemics may have very restricted habitat ranges such as (in the Galapagos) to either the arid lowlands or the moist uplands, when these are present on high islands. In a few groups, the process of evolutionary wing loss manifests in polymorphic stages of gradual loss of hindwings. Subterranean and cavernicolous beetles (see p. 114) may also lose eyes and whereas many such species are clearly related to sighted taxa on the same island(s), Peck (2006) recorded the Galapagos eyeless staphylinid *Pinostygus galapagoensis*, which has no possible sister species on the archipelago. Island beetle faunas commonly include such species whose ancestral origins are not clear. Others can comprise radiations of confamilial groups having separate origins, with several distinct groups of Hawaiian Carabidae, for example, derived from clearly separate colonizations. Thus, Liebherr and Zimmerman (2000) presumed the Hawaiian Psydrini (*Mecyclothorax*) to be a monophyletic radiation, whereas Bembidiini have been derived from repeated colonizations with more restricted radiation after each founding event.

Within the parameters of narrow-range endemism, isolation and vulnerability, many island insects have assumed high profiles in conservation (Howarth & Ramsay 1991), and the islands themselves have sometimes become the major focus for beetle conservation as refuges or sites for introduction from threatened or hard-to-protect mainland species. For example, small rodent-free offshore islands of New Zealand have massive importance for insect conservation in this context. Their endemic species also exemplify conservation need. The large flightless New Zealand weevil *Hadramphus spinipennis* is known only from two small outlying islands in the Chatham group, with historical records from a third, nearby, island. It is associated closely with its major host plant, *Aciphylla dieffenbachii* (Apiaceae), also endemic to the Chatham Islands (Schops *et al.* 1999). *Hadramphus spinipennis* is classified as endangered and *Aciphylla* is a threatened species. The weevil depends on *Aciphylla* for food and mate-finding, so that its conservation depends on the well-being of this host. The plant's habitat is being reduced by increasing areas of shrub and forest on one island (Mangere Island, on which most of the beetles occur), and its distribution is already restricted largely to coastal areas and cliffs on the second (South-East Island). Maintenance of tree-free areas is thereby critical, and Schops *et al.* also suggested the desirability of establishing a third beetle population on another, then unselected, island in the group. *Aciphylla dieffenbachii* occurs on four islands and on Mangere is found as a mosaic of small patches across the island. Patches are occupied by metapopulation segregates of weevils, and individual weevils can walk up to several hundred metres. Mark–recapture studies implied strongly that adult weevils persist in an *Aciphylla* patch until it collapses from over-exploitation, then move to other patches. This metapopulation model was examined by Johst and Schops (2003), with the implication that persistence is possible only at short mean dispersal distances, with substantial mortality during dispersal contributing to low patch connectivity. Thus, in marked contrast to findings of some other metapopulation studies (see p. 64) that emphasize conservation management directed towards decreasing mean inter-patch distances and/or establishing dispersal corridors to increase connectivity, *H. spinipennis* might actually suffer from increased connectivity. A more suitable conservation approach suggested by Johst and Schops (2003) would be to maintain or enhance spatial heterogeneity, for example by increasing the number and size of *Aciphylla* patches without substantially decreasing mean inter-patch distances.

Island habitats

Island habitats are noted frequently in conservation, with the presumption that habitats such as remnant forest patches in largely cleared or alienated landscapes are indeed isolated in a manner similar to that of true islands. For dung beetles in South Africa, Davis *et al.* (2002) found groups of localized taxa on what they designated 'regional islands', areas with a history of ecoclimatic separation from nearby systems. In short, the term 'islands' can be perceived in numerous different ways by biologists, and the extent of realistic parallels with true islands differs accordingly. However, it is commonly unknown whether the presumed alien matrix surrounding these is indeed inhospitable or whether the species of concern can live there, perhaps in small numbers. The problem was discussed for deciduous forest fragments in coniferous boreal forests by As (1993), with these forests the major regional focus of forestry activity. As (1993) suggested that the widespread presence of coarse woody debris rendered these apparently discrete island habitats not isolated for wood-feeding beetles.

   Isolation may be considerably greater in more extreme habitats, with features far different from those of the matrix, and such distinctiveness implies a specialized conservation need for species that are restricted in both distribution and environmental tolerances.

Place: cave beetles

Caves are among the most unusual and isolated island habitats throughout the world and wherever they occur are often associated with an unusual and specialized fauna, including a number of troglobitic beetles. Some of these beetles are both very specialized and highly localized, in some cases restricted to particular caves or cave systems, where isolation has led to speciation. Many are thus of conservation interest as narrowly endemic. Extensive speciation of beetles in cave systems reflects that many of these habitats are indeed highly isolated. Barr (1985) described the eastern North American cave trechine carabids as 'true "living fossils"', in that no possible ancestral stock still exists in the immediate region of their caves. However, as Barr also noted (for Appalachian cave carabids), species status is assessed somewhat arbitrarily, on the basis that this isolation in practice results in distinct biological species. Many taxa are represented, but the predominant groups of troglobitic (obligatorily cave-inhabiting) beetles belong to the Carabidae, Leiodidae and Staphylinidae (including Pselaphinae, sometimes listed as a distinct family). These are indeed the only families of terrestrial Coleoptera reported as troglobites in North America (Peck 1998), although water beetles (Dytiscidae, two species; Dryopidae, one species) also occur in caves. Many such species share the convergent features of being blind and having long legs and pale bodies. Distribution and behaviour can be influenced strongly by food needs and supply (Kane & Poulson 1976).

Speciation in cave beetles can reach substantial levels, in both terrestrial and subaquifer environments. In North America, for example, more than 240 species of *Pseudanophthalmus* (Carabidae) occur in various caves in the eastern USA, among an array of other trechines, as the most abundant group of terrestrial troglobites in the region (Barr 1969; Barr & Holsinger 1985, the latter summarizing much about the evolution of this group). However, many cave beetles, in addition to having very narrow distributions, also occur in only small populations. More generally for troglobitic fauna, Culver (1982) warned of the serious threats that could arise inadvertently by collection of even a few voucher specimens and urged restraint on such activities. In general, as reflected in a list of possible threats to Texas karst cave invertebrates (United States Fish and Wildlife Service 1994), threats to cave fauna can arise from either inside the caves or from the surface environment. These factors include land development (with changes to hydrology, soil erosion, surface species complements, and others), pollution, vandalism (increased local populations, increased visitations) leading to rockfalls or removal of cave material and greater disturbance, and harm from alien species. In Texas, the highly invasive red imported fire ant (*Solenopsis invicta*) has invaded many caves (Reddell 1994). Three troglobitic beetles co-occur in Texas karst caves, and were among the earliest insects listed for protection in the USA. All are endangered and have been found in only a few caves subject to one or more of the above threat categories. Rather than receiving individual treatment, these are discussed together with other notable endemic invertebrates in the caves under a joint recovery plan for endangered karst invertebrates (United States Fish and Wildlife Service 1994). Two of the beetles, *Texamaurops reddelli* (Ketschmarr Cave mold beetle) and *Batrisodes texanus* (Coffin Cave mold beetle) (both Pselaphinae) are among the smallest beetles to receive formal protection, both being only 2–3 mm in length. The two species were initially both treated as *T. reddelli*, with later recognition of *B. texanus* necessitating its removal from the earlier unintentional bulked listing (1988) to a separate listing under its new name (1993). Despite extensive surveys, both species are still known from very restricted areas. The third species, the predatory carabid *Rhadine persephone* (Tooth Cave ground beetle), represents a genus containing more than 60 western North American/Mexican species. Other species of *Batrisodes* and *Rhadine* from caves elsewhere in Texas have also been listed as endangered.

In a rather different context, Australian Dytiscidae have proliferated to produce radiations of species in localized underground calcrete aquifers, and this richness of stygobitic water beetles may have arisen from multiple colonizations (Leys & Watts 2008). More than 80 Australian species, most of them Hydroporinae, have now been described, contributing to what Leys and Watts designated 'the world's most diverse assemblage of subterranean diving beetles', and many of them are small. Smaller beetles (such as *Paroster*, with body lengths 2.5–4 mm) could be adapted better to the interstitial spaces in aquifer substrates and also to a sparse food supply. The beetles typically exhibit specialized troglobitic features, such as loss of eyes and pigment, flattened body shape, reduction of wings and presence of long sensory setae. As noted earlier (see p. 10) many may be narrow-range endemics, by inference from studies on those in aquifers of the Yilgarn Craton, Western Australia, but many other aquifers have not yet been surveyed (Cooper *et al.* 2002). The concept of small isolated subterranean islands for these systems raises important issues for conservation, because the loss or despoliation of any individual aquifer might lead to loss of species that occur nowhere else.

Many of the problems of conserving species of cave beetles are encapsulated when considering two Tasmanian troglobitic carabids, the Ida Bay cave beetle (*Idacarabus troglodytes*) and the blind cave beetle (*Goedetrechus mendumae*), both designated as rare by State authority (Threatened Species Unit 2000a,b). Both species are found only in the Ida Bay karst area of south-eastern Tasmania and have an estimated maximum area of occupancy of less than 8 km^2, with unknown population sizes, no quantified information on declines and little known of any detailed habitat requirements or of any specific threats. The assessment of scarcity and conservation interest thereby rests on small distribution, specialized troglobitic habit, implied threats and need to ensure that the small documented habitats are conserved as the only viable assurance for the beetles' future. Part of the cave system in which these beetles live is potentially vulnerable to operations such as mining and forestry, influencing water flow through the cave system and affecting specific areas within the caves that are regarded as important in maintaining integrity of the cave food webs, as well as perhaps being key components of the beetle habitat. A nearby limestone quarry was formerly believed to pose a similar threat to both species, but this operation has now been closed down. For each species, the sole stated objective of the recovery programme is 'To protect existing Ida Bay Cave Beetle/Blind Cave Beetle habitat from adverse impacts'. Surveys of cave fauna in the area have confirmed the restricted ranges of both species, and recreational access to caves has been restricted, in part voluntarily by cavers instigating protection of a key habitat for *G. mendumae* by using another route to a significant caving area. Research to clarify biology was a key action for both species, with proposals to monitor possible impacts and, for *G. mendumae*, to undertake genetic study to determine the variety of populations in different caves and passages.

Species: beetles with ants and termites

Myrmecophilous and termitophilous beetles live in habitats, ant or termite nests respectively, that are naturally fragmented in the landscape and in an environment created by their host social insect. Many show striking structural or chemical mimicry towards their specific host, and most species seem to be restricted to these specialized habitats, with varying degrees of host specificity. Much relevant background on these beetles was summarized by Holldobler and Wilson (1990), following an earlier bestiary by Kistner (1982), and the complex array of biologies and spectrum of associations is set out there. Suffice, here, to emphasize that associations range from rather casual to obligate and highly specialized, with structural, behavioural and ecological adaptations to fit these beetles for this way of life. The habit has clearly evolved independently many times, and a preponderance of the members of the around 30 families involved are predators, many remaining so even when integrated intimately into their host's colony (Kistner 1982). The taxa involved range from representatives of the smallest beetles (Limulodidae are perhaps wholly myrmecophiles, Ptiliidae) to groups of Carabidae (Paussinae) and representative Scarabaeidae and Chrysomelidae.

In northern Europe, various myrmecophilous beetle species exhibit patterns of local, regional or continental distribution, with some evidence that their richness in nests reflects nest size (volume) and extent of isolation, so that richness (as well as abundance) is resource-related (Paivinen *et al.* 2004). This inference had been made much earlier, with Kolbe (1969) finding that nests of *Formica polyctena* held nine species of beetles, with the largest and most conspicuous nests more hospitable than small ones. More specifically, 16 beetle species were found in nests of the wood ant *Formica aquilonia* in Finland alone, from a literature survey (Paivinen *et al.* 2003), with a broader European total of 75 beetle species associated with 34 ant species. As elsewhere in the world, Staphylinidae (including Pselaphidae, as Pselaphinae) are the predominant group involved. In Denmark, Finland and Sweden the ant species with widest distributions harboured the greatest numbers of beetle species, with a strong positive relationship between distribution of an ant host and its associated myrmecophiles. However, richness decreased northwards, perhaps associated with the ant colonies becoming too sparse to allow the beetles to colonize and persist.

In general, myrmecophiles and termitophiles are among the most poorly understood and poorly surveyed beetles. Numerous species have been described, although many, perhaps particularly termitophiles in the tropics, are regarded as rare, perhaps reflecting the difficulties of retrieval. Many are known there from single localities or hosts, and their precise relationships (as predators, inquilines, or other) with those hosts are unknown. Some taxa, such as Paussinae (myrmecophiles) and Staphylinidae: Aleocharinae (found with both groups of hosts) have very specialized biology. Although so much is unknown about them, these beetles exemplify a conservation situation that is far more widespread and transcends many different guilds and environments, namely that conservation cannot rely on isolated individual study of the focal species alone, but must 'piggyback' on that of the social insect hosts. Knowledge of specific beetle associates may help strengthen the case for this and whilst some hosts are targets for suppression, others may be targets for conservation in their own right.

4

Collecting and Over-collecting

In the past, insect collecting recruited many people to the ranks of coleopterists from their childhood onwards. It is a hobby that has laid the foundations of much of the natural history of beetles, now acknowledged as vital for conservation status assessment and management. However, it now evokes very mixed reactions from people. In part, this is a natural legacy of the common and well-publicized measures of protecting species by formal prohibition of capture, which has led to anyone collecting beetles being viewed with suspicion by well-meaning observers. Almost universally, the formal listing of species in this way for their conservation significance (as protected or threatened species) includes such steps, although in many instances without strong evidence that collecting is indeed a threat. There are clear cases of need for such prohibition, some noted below, but nevertheless beetles have been a popular collectable for well over a century, and continue to be objects of fascination and commercial trade in many parts of the world. As for any other collectable object, rarity or, more broadly, difficulty of obtaining specimens or supposed rarity, begets increased prices and can foster illegal take and trade on the black market. However, the inconspicuousness of many beetles, even if they are rare and of conservation significance, may reduce their attractiveness to some collectors. Hungerford's water beetle (*Brychius hungerfordi*, Haliplidae) is endangered in North America and known from only a few locations in Michigan and Ontario. It is a small (~4 mm long) and rather inconspicuous beetle and, in comparison with larger showier species, is regarded as only minimally threatened by collection. It remains subject to the usual prohibition against take for species under section 9 of the Endangered Species Act, in which take includes 'harassing, harming, pursuing, hunting, shooting, wounding, killing, trapping, capturing, or collecting' the species.

Beetles in Conservation, 1st edition. By T.R. New. Published 2010 by Blackwell Publishing.

Commercial collecting

In contrast, big beetles attract considerable attention. In particular, members of the showier families containing predominantly larger beetles attract attention from collectors and dealers. Perhaps none is more attractive commercially than the larger Lucanidae, and the reported US$100,000 paid for a single specimen by a collector in Japan is perhaps the largest amount ever exchanged for a single dead insect (New 2005). Such trade may indeed foster black-market exploitation of desirable species, with pricing principles akin to those for art-works and rare coins and stamps, so that very strenuous efforts may be made to obtain them, with little or no regard to financial or ecological costs involved. The extent and impacts of poaching to cater for desires of wealthy collectors are difficult to assess. The trade itself is difficult to measure or even detect, and the actual impacts of take on the field populations of target species are usually entirely unknown. Even cursory examination of the variety and quantities of dead beetles advertised for sale on the internet at any time can be a variously depressing or exhilarating experience. For example, the opportunity to purchase 1000 mixed Indonesion Lucanidae at a small per-capita cost, or individual specimens of many larger showy beetles across many families, irrespective of the legal niceties involved for some of these, is understandably appealing to enthusiasts never likely to see such animals in the wild, and the geographical sources for these purchases is considerable. Nevertheless, some numbers quoted (even as estimates) are alarmingly large. In 2001, Japan reputedly imported more than 680,000 stag and rhinoceros beetles for the pet trade, and Goka and Kojima (2004) assessed that trade as 'more than a million beetles imported every year'. The market size of stag beetle trade in Japan was considered to exceed US$100 million (Goka *et al.* 2004).

Some prosecutions for illegal export or trade in listed species have occurred. One important case involved the interception of 1000 individuals of the threatened Lord Howe Island (Australia) stag beetle (*Lamprima insularis*) at Sydney Airport, with the living beetles subsequently returned and released (Leggatt 2003). In general, the collector threat to small, difficult to identify beetles is probably slight, particularly when those groups do not also contain larger species so that the less spectacular forms may then become more desirable commercially to allow enthusiasts to 'complete the set'. Substantial premiums may be demanded for particularly spectacular individuals, for example extra-large or extra-ornamented lucanids, and such trophy individuals are often priced individually or by size category in dealers' lists or on outlets such as the internet. In general, the larger or more elaborated individuals among species that are inherently extremely variable in these features command higher prices. The effects of such selection on field populations are largely unknown.

Listing insect species for protection has sometimes been criticized for drawing attention to the emotional appeal of rarity as a factor increasing the commercial value of the listed taxa, so that rarity equates directly to desirability. By analogy with butterflies, for which more information is available, other human attitudes may also occur. In Australia, the activities of butterfly collectors are the major source of knowledge of a poorly known fauna with few professional lepidopterists, and of clarifying the conservation needs and status of the species of concern (Sands & New 2002). Prohibition of collecting and formal demands for permits, however well intentioned, can alienate these interests, hamper co-operation and be a severe deterrent to others taking up an interest in entomology (Greenslade 1999). However, because many of the species wanted by hobbyists occur in only very small areas or sites, some of these become known as traditional collecting localities for those species. In such places, a succession of responsible independent collectors each taking only a few specimens might collectively have a more serious impact on small populations. In such cases, regulation of collector activity may be justifiable, and understood. For any species of conservation concern, the balance between increased knowledge and possible harm from taking specimens must be evaluated carefully. For *Rhyncogonus* weevils in Hawaii (see p. 110) Samuelson (2003) noted 'The judicious taking of specimens . . . would be justified in surveys that adhere to state and federal guidelines', noting that such future fieldwork on the archipelago is essential to clarify the status of many species and, probably, to reveal further new ones. As Cheesman and Key (2007) commented 'No sensible entomologist would support collecting where this has a negative impact', and often the positive values of fostering information flow may be lost by uncritical prohibitions or other restrictions to bona fide interested people.

However, individual hobbyists and naturalists seeking a few specimens for their personal collections are a very different matter from commercial collectors, whose impacts in seeking every available specimen and lack of long-term care for natural habitats may be much more damaging and whose activities may therefore be a serious threat to sensitive species. These activities may also be extraordinarily difficult to detect or monitor. Lack of distributional and population information on most beetles in the tropics contrasts markedly with, for example, the substantial atlas and database information available for UK species, so that accusations that over-collecting impacts on natural populations are often largely speculative. However, there is no reasonable doubt that serious concerns are warranted in some places. For example, Kameoka and Kiyono (2004) reported that the trade in rhinoceros beetles and stag beetles in Japan includes both native and imported species, and that 90 alien species of Lucanidae were then available for sale in Japan, surveys indicating that these originate from around 25 countries. Indeed, the number of lucanid species that could be imported legally into Japan rose rapidly in response to hobbyist demands, from 34 species in 1999 to a massive 505 species by 2003, enabling Goka *et al.* (2004, p. 67) to write 'We can say that the habitat maintaining the highest biodiversity of stag beetles is Japanese pet shops'.

Perhaps not surprisingly, illegal smuggling from elsewhere in the world accompanied this demand. In some cases, exports of some of the species involved from their countries of origin are formally banned. Goka *et al.* (2004) discussed the case of the widespread *Dorcus antaeus*, with the anachronism that it is absolutely prohibited to collect this beetle in much of its Himalayan range (Bhutan, Nepal, India), but there is no restriction in Japan over its importation. Specimens from this region command much higher prices (around US$5000 per specimen) than those of the same species from Indonesia (around US$50 per specimen).

The high prices available for large specimen beetles for both the pet and deadstock trades encourage direct destruction of trees and forest patches in search of these. Somewhat unusually for a collectable, an individual beetle may initially be sold alive as a pet and after death reach a second interest group as a preserved specimen without loss of commercial value, and indeed may even increase in value. A second related example in the Lucanidae is of the endemic montane species of *Colophon* in South Africa, for which commercial collector interest is recognized as a serious threat, and has led to formal protection measures. However, Geertsema and Owen (2007) claimed that the listing of all species of the genus on Appendix II of CITES (Convention on International Trade in Endangered Species) led to increased attention to the beetles and has increased collecting pressures. The remoteness of many *Colophon* sites renders policing extremely difficult, and largely impracticable.

Three major areas of concern arise from the living beetle trade in Japan (Goka *et al.* 2004; Kameoka & Kyono 2004): (i) collecting and export in countries whose fauna is exploited for this market; (ii) domestic legislation and related control systems in Japan; and (iii) potential impacts of unregulated imports of exotic beetles in Japan, including risks of introducing invasive species, together with introduction and spread of parasites introduced with alien beetles. As with many other pets, release of surplus stock is difficult to control, not least because it is commonly well intentioned. Inadvertent escapes of alien species are also perhaps inevitable considering the scale of pet interest shown. However, interests in Japan also include localized native beetles, stocks of which may be released far from their areas of origin, with presumptive likelihood of interbreeding with local beetles leading to genetic changes. In a few cases, this dilution of population integrity may be enhanced by crossing of native and alien populations, for example of *Dorcus curvidens* and *D. titanus*, both native to Japan but also imported from elsewhere (Goka & Kojima 2004). *Dorcus titanus* comprises 12 subspecies exhibiting island or regional endemism in Japan and is widespread also in much of Asia, with collectors eager to obtain many of the larger forms in particular. Hybrids, at least between some forms, are fertile, and those produced between a Japanese male and a Sumatran female were substantially larger than either parent (Goka *et al.* 2004). A likely consequence, if this trend proves more general, is that collectors will increasingly seek to hybridize stocks to obtain larger specimen individuals. Goka *et al.* also revealed field presence in Japan of mitochondrial DNA haplotypes of exotic forms. The haplotypes of Thailand and Indonesian (Sumatra) forms were specifically noted in wild-caught beetles from different places. Hybridization can thus have the two consequences: loss of regional population integrity and wider loss of national population integrity. Hybridization is perhaps likely to increase with proliferation of imported beetles, so that genetic introgression from field escapes or discards will continue to occur. Although many of the stag beetles imported to Japan are from tropical countries, Goka *et al.* pointed out that many are from higher cooler altitudes, and almost all species pass their long larval life in sheltered environments such as dead trees or soil, so may well be able to thrive outside in Japan. This presumption sets the scene for potential competition with native species from alien species after release or escape.

Bycatch and collector responsibility

Harmful collecting may not always occur intentionally but simply inadvertently or without sufficient thought during scientific sampling or hobbyist activities. Much of the sampling activity for Coleoptera summarized in Chapter 1 relies on selecting the focal species or groups from a much wider array of invertebrates that have been captured unselectively, killed and preserved in pitfall traps, flight interception traps or by other bulk sampling methods. Indeed, for many beetles, it may not be possible to discriminate among taxa without detailed study of dead specimens, but it is inevitable that any such sampling programme yields substantial numbers of bycatch, i.e. non-target organisms that commonly remain unsorted and which are often simply discarded. The alternative of responsible institutional archiving of this material, properly preserved and labelled, is gradually increasing and should be encouraged. However, bycatch may include species of known conservation significance, and any bulk sampling programme can usefully include a literature or expert consultation appraisal for reports of any such species known to occur on the sampling site and their vulnerability to the methods proposed, and random sample inspection to detect any such species with a view to modifying the approach if they are found in numbers thought likely to increase local vulnerability.

Likewise, some direct searching approaches for beetles involve direct destruction or major disturbamce to habitats. These include breaking up dead wood, stripping bark, rolling over logs or stones, sifting waterside litter, and so on, all involving primary disturbance to restricted habitats in search of ecologically specialized beetles. Published codes for collectors counsel against such disturbance, and emphasize the need for moderation in searching, and that materials disturbed or moved should be replaced as accurately as possible in their earlier positions.

5

Alien Species

Effects and interactions with native beetles and other organisms

International importations of beetles for the pet trade (see p. 120) are only a small part of the conservation concern that arises from alien and possibly invasive species. Generally in conservation, alien species (of both plants and animals) are viewed as among the most potent and widespread threats to native biota. Conversely, they may sometimes prove beneficial, for example by augmenting resources, and providing ecosystem services not otherwise available or adequate. Programmes of deliberate introduction of dung beetles into Australia to break down cattle dung, remove pest fly breeding sites and reduce pasture staling can involve numerous species. After the first such introduction in 1968, a cumulative total of around 55 species has been imported and deliberately released, to the extent that some species are abundant and widespread and, in the words of Dadour *et al.* (1999), 'many cattle regions of Australia have become replete with dung beetles'. Any overall adverse impacts may need to be balanced carefully against benefits, including benefits to human welfare, in itself a major driver of deliberate introductions of species to regions outside their natural range. For example, introductions of polyphagous Coccinellidae as potential biological control agents for crop pests (see p. 191) will assuredly remain contentious. They have been revived recently with the spread of the harlequin ladybird *Harmonia axyridis*, first used as a biological control agent in the early 20th century but now known to be an unusually invasive species and regarded as a serious threat to native aphidophagous insects. These threats arise both from competition and the highly polyphagous habits of *Harmonia*, with many of the concerns discussed by Roy and Wajnberg (2008). In hindsight, deliberate introduction of this ladybird to many parts of Europe and elsewhere has proved

---

Beetles in Conservation, 1st edition. By T.R. New. Published 2010 by Blackwell Publishing.

problematical, with rapid spread facilitated by multivoltinism, strong dispersal capability and polyphagy, and resulting in predictions that it will continue to spread and to have increasingly deleterious effects on native coccinellids. It is rare for such widespread invasive effects to occur after deliberate introductions of beetles, but this case demonstrates the need for continuing vigilance and, wherever possible, enhanced practical risk assessment to safeguard against non-target effects.

In short, some deliberate introductions may be benign and others harmful, so that sensible risk assessment must be part of any such new introduction planned. In many places, alien species are widespread, naturalized or otherwise important components of the local economy, such as agricultural or forestry crops, and their interactions with native biota may be both intricate and difficult to assess or predict. As one example, in New Zealand much of the formerly extensive low kanuka (*Kunzea ericoides*) forest on the South Island's Canterbury Plains has been cleared and replaced by plantations of the introduced softwood *Pinus radiata*. A critically endangered endemic carabid, *Holcaspis brevicula*, is thought to have suffered greatly from loss of this native forest and shrubland, to the extent that until recently only two specimens had been known. However, catches from extensive pitfall trapping and direct searching for the beetle in both native forest remnants and a large pine plantation (collectively yielding 47 species of Carabidae over 57,494 trap-days) included five specimens of *H. brevicula* in pitfalls, all of them from pine forest (Brockerhoff *et al.* 2005). All these individuals were from different locations. All pine stands in the region were in at least second rotation, implying that the beetle might survive harvesting disturbance or that recolonization occurs within the 27-year rotation interval. *Holcaspis brevicula* is flightless, but three individuals were found in young pine areas and could have dispersed about 100 m from a mature pine stand. For this beetle, alien softwood *P. radiata* plantations might provide a suitable substitute habitat, whilst the few remnant kanuka patches may be too small to sustain populations (Brockerhoff *et al.* 2005). In New Zealand, as in some other places, plantation forests of alien species are now the only woody cover available over large areas of formerly forested ground. They thus become de facto nominal surrogate habitat for species previously resident in native forests. Comparison of the New Zealand carabids in pine plantations and remnant *Kunzea* patches showed the latter to be particularly important for conservation, but pines to have values considerably greater than alien grasslands in the region (Berndt *et al.* 2008). Pine plantations in Victoria, Australia support substantial numbers of beetle species, with richness in some cases not differing markedly from that of nearby eucalypt forests (Gunther & New 2003). More widely, however, the replacement of native broadleaved forests by exotic conifer plantations has been implicated in many declines of forest beetles, as illustrated earlier.

Dung beetles, ladybirds and pine trees are among the numerous species that have been deliberately introduced to new environments. Many other organisms may arrive without human intention, and these include many serious pests and threatening species. Rodents (mice and rats) are recognized widely as major threats to sensitive native insects, particularly on small islands (see p. 122), and have been implicated as predators in many declines of beetles. For example, Marris (2000) attributed extinction of the carabid beetle *Loxomerus* on Antipodes Island, New Zealand to mouse predation. Mice are the only resident introduced mammalian predators on that island, and the altitudinal distribution of mice was correlated with the distribution of some other beetles. Two species of Tenebrionidae were found alive only at around 300 m, whereas dead remains at around 80 m indicated that these beetles had existed previously also at these lower altitudes. Eradication of mice is one of the most important insect conservation measures on many isolated islands, often to provide a safer environment for translocations or reintroductions.

Alien beetles, usually of a few species in any assemblage, may increase in abundance to constitute high proportions of the individuals present and so dominate assemblages. For example, a survey of the Carabidae in and around Edmonton, Canada yielded 24 species, of which three were introduced (Hartley *et al.* 2007). However, the 21 native species contributed only 491 of the total 3162 beetles trapped, so that the aliens constituted 86% and 79% respectively of beetles in the two vegetation categories sampled. *Pterostichus melanarius* comprised 80% of the total sample. Such species have sometimes been present for a considerable period, with their origins shrouded, and have become naturalized. Correlative surveys of the many European Carabidae thriving in North America and native species suggest that they may have led to reduced populations of some native species, but such inferences made over several decades have proved difficult to address. Concerns devolve on the capability of alien species to extend beyond anthropogenic environments into more natural areas, but there have been few focused attempts to investigate this experimentally. One involved the above-mentioned *P. melanarius*, which although already widely distributed in native deciduous forests in western Canada, was introduced in enclosures to a further stand of aspen/poplar forest to determine how easily it might establish and how it might interact with native species there (Niemela *et al.* 1997). Establishment of breeding populations in the enclosures confirmed the suitability of the introduction sites, and it appeared that native species (six abundant native Carabidae surveyed) did not suffer negative effects in either population size or individual body size from the presence of *P. melanarius*.

The modes of introduction or arrival of most alien beetles, and when such introductions occurred, are conjectural, although methods usually involve natural dispersal, such as flight, passive carriage on air currents, and rafting across water barriers on or in driftwood. Saproxylic beetles may be preadapted to such a transport mode, and could be largely protected from sea water. As Whitehouse (2006) commented, such hitchhikers may be protected by thick bark or among roots, and the longevity of driftwood (with conifer wood floating for periods of up to 10–17 months; Johansen & Hytteborn 2001) may enhance chances of successful dispersal. The number of taxa capable of using this method may be substantial, for example Howden (1977) reported representatives of 31 families of beetles from a few days' survey of beach drift in eastern Australia. Almost any form of importation of goods may be involved in the introduction or alien species, via any available transport mechanism, some of which may even be redundant now. These introductions may arrive as inconspicuous early stages that are particularly difficult for quarantine inspectors to detect, some of whom in any case have sometimes paid rather little attention to insects in the past. The ambiguities were discussed for the 152 introduced species (about 20% of the beetle fauna) of Prince Edward Island, Canada by Majka *et al.* (2006), who recapitulated earlier suggestions (discussed in detail by Spence & Spence 1988) of beetle introductions to North America by (i) inclusion in dry ballast unloaded from ships and (ii) via nursery stock of plants. The first practice ceased after the First World War, but nursery stock importations continue, albeit with soil quarantine measures since 1965. Before then, soil-dwelling invertebrates associated with bedding plants or seedlings might have arrived with little hindrance. Elsewhere, transport by aircraft and importations of timber beetles in wooden goods (including tourist ornaments) and packing crates have occurred frequently. For aircraft arrivals, introductions of insect stowaways are very variable, and influenced by factors such as sanitary care in the country of origin (Caton *et al.* 2006), with flights departing at night also higher risk because of insects attracted to lights of planes being loaded before departure. Amidst the considerable variety of insect species intercepted on arrival at Miami International Airport over a 12-month period, around 10% of incoming flights yielded pest insects, with Scarabaeidae the most intercepted family and Chrysomelidae also well represented. Marine cargo containers made, completely or partially, of wood have for several decades been subject to quality control, through inspection and registration, in Australia. More recently constructed containers are made of steel, but with plywood or timber floors, and in 1996 approximately 770,000 such containers entered Australia (Stanaway *et al.* 2001). Direct searches of about 3000 cargo containers in Queensland in 1996 yielded numerous insects, most of them dead but including a variety of timber pest species, with representatives of five families of Coleoptera. Non-timber pests were represented by members of 11 families of beetles, and included several potentially significant agricultural pests (*Adoretus*, Scarabaeidae, which have been the subjects of biological control attempts in the Pacific region) and stored product pests, as well as predators and fungus-feeders (Stanaway *et al.* 2001). The ecological and taxonomic variety of beetles amenable to such transportation is assuredly large. Export of native insects in containers is also an important consideration, so that (in this case) Australia's quarantine capability and responsibility extends to preventing the export (or further dispersal) of important timber pests such as the eucalypt-feeding cerambycid *Phoracantha semipunctata*, already a significant pest in eucalypt plantations in various parts of the world.

Species transported inside wood are often likely to bypass quarantine, these species including bark beetles and ambrosia beetles of major economic importance as pests (Piel *et al.* 2008), with Scolytinae comprising a very high proportion (93%) of insects intercepted at US ports of entry during 1985–98 (Haack & Cavey 2000). A later appraisal of scolytid introductions to the USA from 1985 to 2000 (Haack 2001) included representatives of 49 genera and beetles from 117 countries, clear evidence of the geographical and taxonomic scale of quarantine needs to assess this one beetle group alone, and one confined largely to the packing materials rather than infesting the goods inside the containers. As of December 2002, 50 species of exotic scolytids were known to be established in continental North America, and Haack considered that more would be found

with additional surveys. Movement is common also on smaller scales, with transfer in either direction between the east and west coasts documented in different
species. Three such harmful species from Russia were among the 43 timber- or
bark-infesting beetle species intercepted on wood imported to Finland in 1985
alone but, as Piel *et al.* (2008) emphasized, continually changing trade patterns
provide rapidly changing opportunities for such alien introductions. Successful
introductions of timber beetles occur when a series of conditions are met, and
these are thereby likely to be a guide to measures to help reduce or prevent
introductions. Those conditions include (i) presence of the invader in the area
of origin of imported timber; (ii) high volumes of trade and geographical
accumulation over time; (iii) absence of pre-export treatments (such as debarking timber, heat treatment, fumigation) to eliminate bark-boring or timber-
boring insects; and (iv) favourable conditions to survive during transport and
later to spread from the point of introduction. In a parallel example in the
southern hemisphere, Brockerhoff *et al.* (2006) analysed interception records
of Scolytinae for New Zealand over the second half of the 20th century,
whilst noting that these were likely to be very incomplete because only a small
proportion of goods were inspected. One additional advantage of analysing pest
groups in this way is that many of the taxa will be recognizable and named.
For the New Zealand series, 1076 of 1505 scolytine records were identified to
species level, and an overall 89.4% to species or genus. They originated in goods
from 59 countries, with many secondary, i.e. imported from countries to which
they had been introduced. However, several of the most frequently intercepted
species (including the two most frequent, *Hylurgops pallictus* and *Pityogenes
chalcographus*) were identified only in cargos from their natural areas. Collectively,
103 species across 38 genera were recorded, with most of the more frequent
taxa associated primarily with conifer hosts and including some important forest pest species. Detection of known significant species, such as the Asian pest
cerambycid *Anoplophora glabripennis* in North America, may trigger increased
vigilance. Some ambrosia beetles (also Scolytinae) have an incestuous breeding system, involving mating between siblings, so they may be especially good
colonizers in being largely immune to genetic erosion resulting from small
founding populations. Three such species are among relatively few beetles to
have successfully colonized tropical forest (Central America) (Kirkendall &
Odegaard 2007). However, in numerical terms, Scolytinae were substantially
second to Curculionidae in the North American interceptions (Haack 2001),
with 6992 records compared with 42,915 weevils. Most of the latter are not
associated with wood, and reflect the high variety of plant materials and deriva-
tives in imported goods.

Most alien Coleoptera are not truly invasive, and may be regarded more
accurately as naturalized. Introduced Carabidae in Canada, for example, are
predominantly species of disturbed or anthropogenic environments in Europe,
and have mostly remained in similar habitats after arrival (Spence & Spence
1988), rather than invade climax native boreal forests. They have not extended
their ecological range markedly. However, a few have indeed spread, and these
include short-winged taxa incapable of flight. Chance human-aided transport
may combine with natural dispersal to produce such patterns. For *Pterostichus*

melanarius (above), the most widespread introduced carabid in western Canada, Spence and Spence noted peripheral populations of short-winged individuals, whilst long-winged individuals were retained in populations near coastal centres where they are presumed to have been introduced. First reported near Edmonton in 1959, later captures (together with those of native carabids) and study of its range expansion (Niemela & Spence 1999) led to two important inferences: (i) that *P. melanarius* catches correlated positively with those of native species, so that it does not have any strong or consistent adverse effects on native species; and (ii) that native species are not able to prevent invasion of forests by *P. melanarius*.

Two patterns of post-arrival dispersal are common among insects, and are exemplified by the spread of a buprestid (*Agrilus planipennis*; see p. 189) since its original discovery in Ontario and Michigan in 2002. It was presumed to have been introduced in wooden packing material, and has spread by both gradual range extension (aided by local flight) and by longer-distance transport by people, in firewood or in infested ash saplings (Muirhead *et al.* 2006).

By whatever means, this species and numerous other beetles representing almost all trophic levels have arrived, colonized, spread and become permanently established in new environments. Whether valued as biological control agents, condemned as pests or with largely unheralded (and possibly wholly innocuous) roles and interactions with species already present, the progress of many such species merits careful monitoring and appraisal. The lily leaf beetle *Lilioceris lilii* (Chrysomelidae) was first detected in North America in Quebec in 1943, presumed to have been imported accidentally on Asiatic lilies. Since 1978 it has spread substantially to occupy much of eastern Canada and the nearby USA. Concerns have arisen as to whether this range expansion might be accompanied by the beetle (a pest of ornamental lilies) expanding its food spectrum to threaten any of the 21 indigenous lily species (Bouchard *et al.* 2007), several of which are already of conservation concern. In this case, *Lilioceris* can be recognized easily, as there are no similar native beetles in North America.

Harm to native plants may result from introductions of herbivores as biological control agents against exotic weeds, despite screening these agents for safety in advance. The European weevil *Rhinocyllus conicus*, for example, was introduced to North America as a potential control agent for *Carduus* thistles, and another weevil feeding on thistles (the Eurasian *Larinus planus*) is adventitious there. Interactions of both these species with non-target native thistles have posed problems (Louda *et al.* 2003), and indicate the complexities of predicting the impacts of such alien species, whether introduced deliberately or by accident and even following extensive screening beforehand. Two possible syndromes contribute to perceived changes in host plant range by insect herbivores in a new area: (i) the species itself changes, under some form of 'ecological release', to expand its feeding spectrum and to acquire novel plants not previously available to it but which it could not accept at the time it was screened for specificity; or (ii) the initial screening was not sufficiently critical or comprehensive. Extensive testing of *R. conicus* against North American *Cirsium* suggested that the weevil's effects on these would be negligible (Zwoelfer & Harris 1984). However, after release it indeed attacked a number of native species, and Louda (1998) suggested that some important ecological variables were not considered sufficiently in the pre-release testing. The cascade effects of *R. conicus* extend to its implication in the decline of a native tephritid fly (*Paracantha culta*). With increased numbers of the weevil in the flowerhead guild of Platte thistle (*Cirsium canescens*), change in use of this resource reduced numbers of the fly emerging (Louda *et al.* 2003). *Larinus planus* has been distributed in western North America as a control agent for Canada thistle (*Cirsium arvense*), but has had more impact in Colorado on a sparse native taxon (Tracy's thistle, *C. undulatum* var. *tracyi*). This weevil had earlier been rejected for deliberate introduction because of its broad host range but, following its discovery in the wild in the USA, was partly re-evaluated and deemed safe for wider distribution. The background to these two cases (Louda *et al.* 2003), with *R. conicus* known to develop on more than one-third of the more than 90 native North American *Cirsium* species, clearly indicates the difficulties of predicting the outcome of species introductions, even for species that have been reasonably well studied. In such cases, native relatives of known food species may prove to be particularly vulnerable, as has long been recognized by biological control practitioners.

For almost any invasive alien species, it is difficult to wholly exclude the possibility that they might threaten native species, simply because of those ecological characteristics (such as dispersal capability, high fecundity and competitive ability) that facilitate invasion. In many instances, any such threat is likely to be contributory rather than primary, but even a small additional 'threat load' to already threatened species might prove critical. Thus, spread of the Argentine ant *Linepithima humile* along permanent streams in California is considered likely to have a significant impact on the long-term persistence of the threatened Valley elderberry longhorn beetle (*Desmocerus californicus dimorphus*, Cerambycidae), which occurs only on blue elderberry (*Sambucus mexicana*) (Huxel 2000). One of the three critical habitat areas designated for the beetle has the ant spreading towards the headwaters and, in this context, Huxel suggested that the threat to *Desmocerus* from loss (of > 97%) and fragmentation of riparian woodlands may be augmented by the Argentine ant, with this supposition supported by the ant's established record of impacts on native species in fragmented habitats.

Alien species may often be distinguished easily among relatively well-known faunas, but predicting and documenting their spread and interactions with native biota is often difficult, even in small geographical areas. Three species of Carabidae: Platynini are regarded as adventitious in Hawaii (Liebherr & Zimmerman 2000), within an array of 129 species of this tribe known there.

1 *Calathus (Neocalathus) ruficollis ruficollis*, known from only seven specimens and reported only from the Honolulu–Pearl Harbor area of southern Oahu.
2 *Laemostenus (Laemostenus) complanatus*, probably introduced to Hawai'i Island in the 1940s. It has since been reported from several localities on that island and from a single record on Oahu and may have potential to compete with native species in native forests and forest edges.
3 *Metacolpodes buchanani*, known from the archipelago only since 1991, but already reported from the five major high islands. This species has high

fecundity and disperses well (e.g. many of the specimens were taken at light), and Liebherr and Zimmerman (2000) characterized it as an exotic tramp species with potential to compete with many native platynines.

Native *Mecyclothorax* carabids in Hawaii have been affected by the arrival of the adventive European carabid *Trechus obtusus* in Maui in 1999, with some populations having declined in abundance (Liebherr & Kruschelnycky 2007).

Alien beetles as vectors

Beetles may not travel alone. They are commonly accompanied by other organisms – parasites, diseases or a wide variety of symbionts or mutualists whose ecological impacts are largely unknown, but which have the potential to be introduced successfully into the new area and, in some cases, to spread from their vector species to others. Parasites or, more generally, phoretic associates of most beetles have been little studied, for example the mites frequently found (sometimes in considerable numbers) on the bodies of scarabaeoids are poorly documented both taxonomically and ecologically. Many species of mites have been found attached to living stag beetles imported to Japan and at least one is considered to be pathogenic (Goka *et al.* 2004). Alien mites introduced with this pet trade appear to have not yet established wild populations in Japan, although host switching has been observed in pet shops (Okabe & Goka 2008). Large numbers of beetles infested with canestriniid mites occur in pet shops. The mites live on the beetle's body, commonly being found under the elytra, and restricted to adult beetles. They are presumed to be commensal, feeding on exudates, but little detailed information is available on their biology. The mites seem to thrive and increase in numbers in the controlled conditions of air-conditioned shops, and move easily to other beetles, from where they could reach the wild (see p. 120).

   A rather different emphasis of vector importance involves beetles that can transmit plant diseases. One of the more notorious cases involves the scolytids *Scolytus multistriatus* and *S. pygmaeus*, both important vectors of Dutch elm disease, a vascular wilt disease caused by two closely related species of Ascomycota and which has devastated elm trees (*Ulmus* spp.) in many parts of the world. Both are native European bark beetles, but *S. multistriatus* in particular has been introduced into North America and elsewhere, where its impacts include transmitting the disease from diseased to healthy trees by carrying the ascospores and conidia of the fungus (*Ophiostoma novo-ulmi*) on the body surface. These scolytids are also vectors for a considerable variety of mites: in a survey of adult beetles in Austria, nine species of mites were recovered, together with two species of nematode worms (Moser *et al.* 2005). Their biology and roles are largely unknown, but unwitting introductions of such variety may pose concerns, in addition to the disease vector role alone. Nevertheless, it is clear that these two scolytids may not be in any way exceptional in carrying a variety of phoretic mites, and suites of species have been reported also from other scolytids and many other beetles. It is possible that some might be additional vectors of Dutch elm disease, or of other fungal diseases of host plants. Other than for a few species of these mites (notably *Pyemotes scolyti*, the most frequently taken species in this study, and others), little biological knowledge is available, but a variety of trophic roles are represented, as in the better-studied mites associated with conifer bark beetles. Among these, particular mites are parasitoids of beetles, predators of insects or nematodes, omnivores, or specialized fungus-feeders.

More generally, bark beetle–fungus interrelationships can be very complex, with the variety of beetle-feeding habits associated with widespread dissemination of fungi from tree to tree. Webber and Gibbs (1989) distinguished two rather different categories of vector beetle–fungus associations: (i) aggressively pathogenic fungi disseminated by several, in most cases closely related, species of beetles; and (ii) particular bark beetle (or other) species that undertake mass attacks on trees and which may transmit several species of fungus, mainly of the 'vascular stain' category. Because of the economic impact of these diseases, considerable research has been undertaken to clarify insect–fungus associations, but details of many remain unsolved. The diversity of possible organisms associated with beetles is exemplified also by the saproxylic bess beetles (Passalidae). Although these are most commonly not aliens, Lichtwardt *et al.* (1999) commented that few insects have 'so varied an assemblage of parasitic and commensalistic organisms' and cited references to accounts that include mites, nematodes, protozoans and flies within this array, in addition to numerous fungi, with these carried by both the beetles and their mites. It can be presumed that virtually any uncritical transfer of living beetles from one part of the world to another has potential to distribute a number of associated species, many of unknown impacts in new areas. As for alien species of any animal or plant group, in such cases of uncertainty the precautionary principle ('If in doubt, keep it out') is a useful maxim to follow.

6

Pollution and Climate Change

Pollution

Two major categories of pollution (broadly, the discarded by-products of human activities, most commonly applied to effects of chemicals released into the environment) are among the threats enumerated for beetles. The first, industrial pollution, encompasses many forms, but is perhaps most obvious as chemical changes to water bodies, soil or the atmosphere by accidental spills or planned discharges of chemicals from human activities. It also includes some other changes to the environment, for example release of heated water from industrial cooling towers or factories. The second, pesticides, is largely a consequence of intensive agriculture and forestry, perhaps particularly from crop protection and fertilizer applications. Insecticides are a particular concern in insect conservation because, by definition, they are meant to kill insects! Many pest beetles, some related closely to species of conservation concern, may be targeted for insecticide applications to suppress their impacts. Concerns from pesticides fall predominantly into two major groups: (i) non-target effects in the areas of application, and (ii) drift, whereby they are carried in various ways to other environments, usually abutting those to which they were applied but occasionally more distant. Residues may, through transport or incorporation into food webs, convey later non-target effects over larger scales. Collectively, they have a considerable variety of lethal and sublethal effects, resulting in declines of vulnerable species and changes in assemblage composition. Control of insecticide drift may at times become an important component of conservation management. Omland (2004) noted substantial mortality to the endangered Puritan tiger beetle (*Cicindela puritana*, see p. 141) as a result of overnight insecticide drift, detected by sudden loss of marked beetles.

Beetles in Conservation, 1st edition. By T.R. New. Published 2010 by Blackwell Publishing.

The effects of pesticides on non-target beetles have been explored very inadequately, except as subsidiary to research on 'safe' pesticide use against particular pest beetles in forests, orchards and field and ornamental crops, but scales of influence range from local to much more widespread. Much of the early relevant literature was summarized by Heliovaara and Vaisanen (1993). The approach in a number of studies (such as Huusela-Veisola 1996) has been to compare assemblages, mainly of carabids, in field experiments using pesticide-treated and pesticide-free paired field plots, or different pesticide applications in the same crop area. Both scale and timing of applications may be influential. In the above-cited study, numbers of Carabidae in pitfall traps were substantially lower in dimethoate-treated plots than in untreated plots, but the effects were quite short-lived, lasting only about 4 weeks, after which beetle numbers increased, probably by immigration. In Huusela-Veisola's study, the application was made in spring, coinciding with a major peak of beetle activity. Beetles active later in the year would generally not be affected, although some pesticides may have residual effects.

Earlier, Freitag (1979) reviewed the effects of pollution on carabid beetles, and considered their value as sensitive responders to both industrial pollutants and agricultural pesticides. Industrial accidents can never be anticipated fully of course, and Freitag referred to Evans' (1970) appraisal of an oil spill off the coast of Santa Barbara, California in 1969, when oil deposits on shore apparently caused extirpation of some populations of the carabid *Thalassotrechus barbarae*, which occurs only along part of the Californian coast. Freitag also enumerated the effects of pollution on carabid distributions from (i) discharges of sulphur compounds from a factory in Poland and from a paper factory in Ontario (both associated with gradients of beetle abundance with distance from source); (ii) traffic pollution (with some carabids having higher body lead content near the edges of a busy road than when collected from meadows); and (iii) pesticides.

The general term 'pesticides' covers an enormous range of chemicals and contexts, with many putative non-target effects needing much more clarification. For example, application of veterinary anthelmintic drugs to domestic stock is sometimes of significant concern. Up to 90% of the given dose of avermectins (with ivermectin and abamectin as the two major formulations used widely) is excreted in animal faeces and persists there to affect dung-feeding insects (Strong 1992; Floate 1998). Their effects include increasing the developmental period, reducing larval survival and reducing reproduction of developing adult beetles. A single standard injection of ivermectin to cattle had substantial influence on the South African dung beetle *Euoniticellus intermedius* in comparison with emergences from dung of untreated cattle; emergences of *Onitis alexis* were also delayed (Kruger & Scholtz 1997). These authors suggested that potential risks to local dung beetle faunas might be reduced by treating the cattle at times outside the main beetle breeding periods, a recommendation also made by Ridsdill-Smith (1988) in Australia. For some Mediterranean dung beetles, including large burrowers that dig deep burrows and bury the dung (e.g. some species of *Copris, Geotrupes, Onitis*), the dung may then decompose more slowly than when on the surface, so that risks to beetles persist (Lumaret *et al.* 1993). Rate of breakdown of dung can be used as an index of avermectin influence, and several of the studies noted by Dadour *et al.* (1999) have described such effects or, conversely, their absence. Dung containing ivermectin residues is less attractive than uncontaminated dung to *Onthophagus australis* in Australia, so that contaminated dung is not broken down as rapidly by beetles.

Climate change

The reality of climate change and that it has severe implications for the distribution and well-being of many plant and animal species is now widely accepted, supported by considerable evidence of changing geographical ranges and seasonal appearances of insects in apparent response to increased temperatures. The evidence is strongest for well-known groups such as butterflies and dragonflies in northern temperate regions, where changes can be appraised against a strongly documented historical background, and related to time intervals. For example, poleward shifts in distribution, either as range expansion or accompanied by loss of occupancy along the trailing edge, have been reported in British fauna of these groups (Warren *et al.* 2001; Hickling *et al.* 2005). Surveys of several British beetle groups (Carabidae, Cerambycidae, Cantharoidea and their allies) all revealed northward latitudinal range shifts over a 25-year period (Hickling *et al.* 2006). Such latitudinal movements are paralleled ecologically by shifts of species from lowlands to higher altitudes, both patterns reflecting distribution along geographical gradients along which climate changes may be gradual. Direct observations and meta-analyses attest to the reality of these changes, and there is no reason to suppose that they are not paralleled in all other insect groups, and that the consequences are largely unknown. In addition to distribution changes, changes in phenology have been described, so that many highly seasonal temperate region butterflies now appear earlier in the year than they did a few decades ago. Increased temperatures thereby affect both site suitability and synchrony with important resources. An earlier study of the Adonis blue butterfly *Lysandra bellargus* in southern Britain (Thomas 1983) noted its strong limitation to south-facing slopes with short swards and bare ground, reflecting their need for heat, with this limiting their distribution and being a key need for conservation at the northern margin of its European distribution. Several such species now have wider distributions as more northerly areas have become suitable for occupation, and their conservation management has been modified accordingly.

The consequences for British insects as climate change proceeds are that alien species or migrants from continental Europe may both increase in number and find it easier to establish (Sparks *et al.* 2007) and perhaps be brought into increased competition with resident insects under stress on the southern fringe of their range, possibly in turn influencing their displacement northwards as conditions there become suitable. However, as Alexander (2003) put it for Cantharidae and Buprestidae in Britain, 'habitat loss and degradation remain the main issues for conservation attention. Global climate change will merely exacerbate the problems'. Nevertheless, loss of resources and sites due to changing climate, and the uncertainty of their juxtaposition elsewhere in the future, are clearly of concern.

Wilson *et al.* (2007) summarized similar scenarios in many insect groups: increased risk of extinction at lower or cooler sites; increased chances of losing synchronicity with resources such as particular plant foods; and changing exposure to seasonally active natural enemies, including nesting insectivorous birds, that might be affected differently by the same level of temperature or other change. Temperature changes also influence insect growth rates, fecundity and many other aspects of individual or population fitness, and the capability of many insects to adapt to such imposed changes is largely unknown. Diapause regimes may be changed, affecting fundamental seasonal timing of development patterns. However, the groups of insects suggested to be especially susceptible to climate changes overlap largely with those of conservation concern, among which similar features may predispose them to extinction. These include (drawing from Peters & Darling 1985) ecological specialists with little adaptive capability or ecological flexibility; narrow-range endemics; species already found at the climatic extreme of a possible range (such as alpine species near the upper points of an altitude range, and which so have nowhere to go); poor dispersers; small populations; genetically impoverished populations; and populations with peripheral or disjunct distributions.

Numerous beetles fit one or more of these categories, but most of those studied on single sites or otherwise narrow distributions have been subject only to site-based conservation measures designed to enable them to persist and thrive in the places where they already occur. Their ability to actively change their ranges, or to adapt to changed climate parameters *in situ*, is usually completely unknown. Conservation measures should ideally acknowledge that, however effective they may be at present, the sites on which they are focused may become entirely unsuitable for the focal species in the future. A projected need to move, or to be moved (New 2008; see p. 138), emphasizes that suitable habitats must not become too fragmented in the landscape (Hanski & Poyry 2007) and the value of planning conservation along relevant environmental gradients to facilitate chances of more gradual adaptations by insects.

Rather more attention has understandably been paid to the actual or projected changing distributions of pest attack by particular beetles, especially some associated with forest damage. Predicting movements of invasive alien species (see p. 127) in relation to climate tolerances is a closely allied parallel need. In both contexts, as in conservation, consideration of possible ecological gradients may be a central theme as a species' climate envelope becomes gradually changed on a current residential site. Modelling the temperature and humidity needs of a species has been a valuable tool in helping to predict the success of introduced biological control agents, such as the dung beetles introduced to Australia to help degrade cattle dung (see p. 122), and in demonstrating the relative suitability of different species across a wide area of the country. However, simple 'climate matching' may be only a general guide to suitability, with species-specific developmental parameters adding greatly to predictive value, as shown for the mountain pine beetle *Dendroctonus ponderosae* by Logan and Powell (2001). For such economically important beetles, the substantial costs of obtaining this additional information are justified. In contrast for the many species with narrower appeal to people, including most species of conservation interest, the exercise may be difficult to undertake. The analysis of *D. ponderosae* indicated that seasonal adaptability could be countered by the inflexibility imposed by diapause, but greater adaptive ability may accompany direct temperature control of seasonality. In this species, synchronous development is of critical importance, because the beetle operates by mass attacks on trees. Both timing and synchronicity are central needs in the univoltine life cycle, and some details are noted here simply to indicate the likely responses to climate change even of abundant beetle species. The adaptive need for *D. ponderosae* is for adults to emerge early enough in the season to allow maximum time for oviposition but late enough to avoid the lethal effects of spring/early summer temperatures (Logan & Powell 2001), so that optimal adaptive timing reflects a balance between maximizing reproductive output and minimizing mortality. The need for synchronicity reflects the mass attack lifestyle – the beetles must kill the pine tree host in order to reproduce, and the beetles can overcome the host's substantial chemical defences only by weight of numbers, contingent on large numbers of beetles being available at the same time. Details are given by Raffa and Berryman (1987), who showed how fitness increases initially as beetle numbers rise, and then declines with higher densities leading to increasing competition for phloem. Temperature changes, or range changes resulting from these, may have complex influences, and for almost all beetles these are completely unknown.

Climate change is not, of course, purely a modern phenomenon and the Quaternary fossil history of beetles reveals many examples of substantial range shifts. As one example, detailed by Ashworth (2001), the currently subarctic carabid *Diacheila arctica*, found in the northern parts of Scandinavia and adjacent parts of Russia, occurred in Britain during the last glaciation. It became extinct there with rapid warming about 10,000 years ago, after invading and thriving for some 1200 years. Historically, it seems that beetles tended to respond to climate change during the Quaternary predominantly by dispersal, eventually leading to large-scale changes in distribution (Ashworth 2001). Ashworth noted, however, that it is less certain whether dispersal will ensure their survival in the future, because of the outcomes of fragmentation of natural habitats reducing their chances of colonizing new habitats. Documented beetle extinctions are rare in the Quaternary (see p. 5), but many species are probably now far more vulnerable than at any previous time in their evolutionary history. Paralleling Alexander's comment above, Ashworth (2001) noted that 'Global warming potentially makes a bad situation worse'. Acknowledging this, Ashworth (1996) suggested that knowledge of past responses of beetles should be considered when management protocols for conservation are developed, and noted several aspects of responses to future climate change, drawing on information from the fossil record of arctic Carabidae but relevant also to many other contexts.

1   Geographical ranges of species will change, perhaps rapidly, and arctic Carabidae may be useful as primary monitors of climate change.
2   Physiological adaptations will accompany changes in distribution but are unlikely to lead to speciation.

3   Landscape barriers, resulting from fragmentation, will lead to regional extinctions, and alpine species will be lost by increase in forests and upward movement of the treeline.
4   Species extinctions will occur also in other habitats in which dispersal is limited by human activities.

Life-history flexibility is an important facet of adaptive ability to different temperatures, and may differ across members of an assemblage at different altitudes (Carabidae; Butterfield 1996). However, whilst temperature may indeed be a determinant of carabid distribution and activity, the flexibility of their developmental patterns may also be limited by other factors. These beetles need to have sharp mandibles in order to feed effectively, and Butterfield noted that seasonal development may be adapted to conserve mandibles from wear, particularly so that the beetles can overwinter and feed successfully the following season. Likewise, climate variables are important determinants of species richness in scarabaeine dung beetle assemblages, with temperature perhaps particularly significant (Lobo *et al*. 2002, for France) and relatively few species in some local faunas able to tolerate cooler temperatures. This scenario differs somewhat for Aphodiinae in the same region (Lobo *et al*. 2004), as many of these beetles are well adapted to cooler conditions.

7

Components of Beetle Species Conservation: *Ex Situ* Conservation

Most practical beetle conservation is entirely field-based, most commonly focusing on particular sites where target species occur or on the particular resources they need, and some of the approaches to field population management are discussed in the next chapter. However, some cases have demanded a broader approach, in which so-called *ex situ* conservation measures are also involved. For some threatened species, such as the North American burying beetle *Nicrophorus americanus*, these are formally stated components of recovery plans, and then go hand in hand with field management measures. In some other plans, such topics as translocations are mentioned as reserve or subsidiary management options. Recovery plans for many threatened beetles, in many places, include consideration of translocations or reintroductions as part of the portfolio of possible measures.

Ex situ conservation

It is not uncommon in insect conservation programmes to move insects in the field to augment small existing populations or to establish new populations. In some cases this exercise is not direct translocation but involves periods of holding in captivity, or more extensive captive breeding exercises, so that conservation may be based strongly on *ex situ* components of a plan. In some extreme cases, awareness that the sole known site of a species or significant population is to be sacrificed for development, or has been degraded to the extent that it may no longer sustain a viable population, ensures that conventional *in situ* (on-site) conservation approaches are no longer viable. The major practical option available may then be to remove as many individuals as possible, as a salvage operation.

More broadly, three main strategies are commonly canvassed.

Beetles in Conservation, 1st edition. By T.R. New. Published 2010 by Blackwell Publishing.

1   The need to augment an existing field population by introducing individuals from other populations or from captive stock.
2   The formation of new populations on currently unoccupied sites in order to increase the species' range or increase the overall number of populations as a counter to extinctions.
3   The rescue of individuals, as above, when mortality is high on the site or the site is to be destroyed. This form of salvage is extreme, particularly if it is not known how the species may fare in captivity, and as much prior preparation as possible should be undertaken.

Each of these may involve substantial practical difficulties as an original and putatively one-off exercise. Any form of introduction or reintroduction of a beetle must be a carefully planned exercise with basic considerations including (i) effects on the donor population through removal of individuals; (ii) whether to translocate and release removed individuals or to establish a captive breeding stock to build up numbers for later use, or both; and (iii) when, how and where to liberate individuals in the future, with due regard to selection and preparation of receptor sites, and the need to document the whole operation and monitor the outcome. Each stage, as in the sequence proposed by Meads (1995) (Fig. 7.1), may need original appraisal. The exercise is usually far from simple and, in many cases, site preparation needed to host a new population or support greater numbers may be a complex and long-term operation involving, for example, plantings and engineering works to modify drainage or topography and perhaps lengthy negotiations for site purchase or change of tenure to assure future security.

As a general working rule, it has traditionally been recommended that translocations for augmentation or for establishment of new populations take place only within the documented historical range of the species involved. However, taking into account the implications of climate change on the current distributional ranges of many insects, this perspective may change rapidly (New 2008) to also encompass sites likely to become suitable in the future.

In general, rather more information is available on Lepidoptera and Orthoptera translocations and reintroductions than specifically on beetles, but the principles are universal in fostering responsible practice and maximizing chances of survival through populations that are sufficiently large, secure and genetically variable to be sustainable. As for *in situ* conservation, knowledge of a species' population form, resource needs and phenology are important in planning, and in preparation of receptor sites. Any conservation strategy may need to incorporate consideration of sensitive stages in the life cycle, for example as Matern *et al.* (2008) demonstrated for the long-lived *Carabus variolosus* in Germany (see p. 67).

For many species of conservation concern, possible donor populations may themselves be small and founder stock for new populations thereby derived from few individuals and be genetically restricted, so that future inbreeding or other deleterious effects may become a consideration, particularly should long-term captive breeding (of more than two to three generations) be planned. Any *ex situ* conservation attempt is likely to be a complex and protracted exercise and cannot be assumed to be a quick fix by simply moving beetles from one place to another. Occasionally, this approach is feasible (as with the field cricket *Gryllus campestris* in Germany; Hochkirch *et al.* 2007) but for many ecologically specialized species the three categories of influential factors noted by Hochkirch *et al.* (namely ecological background, the protocol for translocation, and the level of scientific and administrative support available) may need very careful appraisal beforehand.

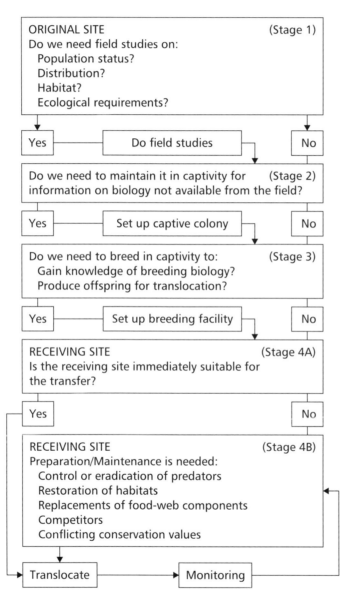

Fig. 7.1 Various sequential decisions to be made in evaluating a possible insect translocation strategy based on a scheme designed for New Zealand weta. (After Meads 1995 with permission.)

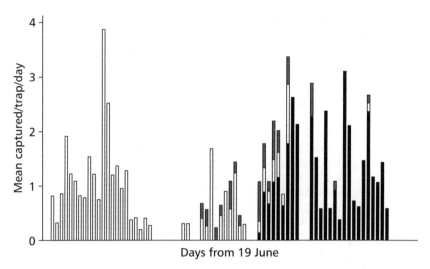

Fig. 7.2 Seasonal incidence of three age classes of *Nicrophorus americanus* captured by baited traps in Nebraska, USA. Daily captures are shown, commencing on 19 June. Open bars, mature beetles; black bars, teneral beetles; hatched bars, senescent beetles. (From Bedick *et al*. 1999 with permission.)

Even issues such as when to collect founder individuals may be important. Figure 7.2 shows the seasonal incidence of three major age groups (teneral, mature, senescent) of adult *Nicrophorus americanus* from a survey in Nebraska (Bedick *et al*. 1999). These categories are separable visually: (i) teneral beetles have bright markings, and the exoskeleton is soft and translucent, with elytral pubescence pronounced; (ii) in mature individuals (second season adults) the pronotal markings are darker than those on the elytra; and (iii) senescent beetles have pale elytral markings and commonly show signs of wear, such as chips or scars. Only mature beetles were captured early in the season, with later transition to other categories. To establish breeding stocks, mature individuals should be used, because senescent individuals may be post-reproductive. Consequently, founder stock must be collected during the early part of the summer.

New populations

Establishment of new populations is a common aim stated in recovery plans, but rather few cases have been documented carefully for beetles. Experimental translocations of the threatened tiger beetle *Cicindela dorsalis dorsalis* in North America clarified how this might be achieved (Knisley *et al*. 2005) to fulfil one of the major objectives of the formal recovery plan. In common with some other North American tiger beetles, substantial declines of *C. d. dorsalis* had followed disruption of sandy beach habitats for human recreation, so that there was abundant opportunity to consider reintroductions to sites within the beetle's historical range. Early attempts to translocate adult beetles were not successful. Adults were marked (by removal of pronotal setae) but post-release surveys indicated that they rapidly dispersed from the release sites and did not persist to breed there. Translocations of larvae were considered more likely to succeed, and the two beaches selected as translocation sites were not intensively used by people because of distant access, and the beaches were closed to public access from April to September to protect nesting piping plovers (*Charadrius melodus*, a federally listed bird), a period also important for the beetle's development. The five donor sites all had relatively large beetle populations (> 1500 adults), and larvae were released individually, so that they could also be monitored individually on small patches of beach above high tide level. Larvae commenced to dig burrows within 60–90 minutes of release, and the exercise was repeated over several years from 1997 on. Later counts of adults showed an increase for several years but then declines from a high of 749 in 2001 to only six in 2004. One suggested reason for this decline was predation by gulls. Despite the failure of this exercise, Knisley *et al.* believed that the experience was an instructive foundation for further translocation studies.

A decision about whether and how to translocate a threatened beetle may have far-reaching consequences, and these may be difficult to anticipate or balance. It is often difficult to provide an objective risk analysis of the various options because of poor biological understanding, but one informative attempt involved a population viability analysis of the Puritan tiger beetle (*Cicindela puritana*) in Connecticut. There, the beetle has been reduced to two metapopulations, one of them very small (Omland 2002, 2004). One objective of the species' recovery plan is to increase both the abundance of individuals and the number of metapopulations. The suitable river beach habitat is very patchy and although adults can fly, they apparently only rarely move between metapopulation segregates. Omland (2004) projected six tactics as possible constituents in conservation management, and assessed the possible risks of each by population simulations. The broad scenario (Fig. 7.3, in which 'river distance' reflects the location of the major metapopulation between kilometres 46 and 60) involved aspects of translocation, habitat management and reduction of mortality, with the objectives of the analysis being to assess (i) which of these might pose minimal or least risk, (ii) whether individuals for translocation may be removed without risk, (iii) whether augmentation or attempt to create a new population might be preferable, and (iv) which tactic poses least risk of regional extinction. The decisions therefore involve (see Fig. 7.3):

1   comparison of tactics 0–3;
2   whether tactic pre-4 or pre-5 would more exacerbate risk of decline;
3   whether tactic 4 or 5 would yield greater reward;
4   which among all would maximize the chances of having two to three large metapopulations in the future.

Because the extent of suitable habitat decreases as vegetation succession progresses, clearing vegetation (by cutting and subsequent herbicide treatment of stumps) was viewed as a way to increase site carrying capacity, with the assumption that this could be doubled at kilometre 49. Translocation was envisaged as removing 100 larvae from kilometre 50 in each of two years, and comparing results from

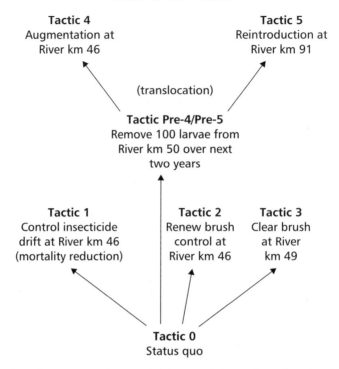

Fig. 7.3 Options for conservation management of the Puritan tiger beetle *Cicindela puritana* in Connecticut, USA. (From Omland 2004 with permission.)

augmentation of an existing population unit and establishing another. Modelling revealed tactic 3 (Fig. 7.3) to be best for minimizing risk of decline, with tactic 1 the next best, followed by tactic 2. Tactic 5 was more likely to succeed than tactic 4. For this species, the modelling exercise proved a useful tool in setting priorities among various conservation options, including translocations.

Salvage or rescue operations

The flightless Frégate island giant tenebrionid *Polposipus herculeanus* (Tenebrionidae) was known only from this small island (of around 200 ha) in the Seychelles. It had been reduced to very low numbers in the wild by the late 1990s by predation from rats since they were introduced accidentally in 1995. In 1996, individuals were captured to found a captive population at the London Zoo, with the aim of conserving the species through captive breeding and the intention to eventually re-release beetles to the wild. Pearce-Kelly *et al.* (2007) reported that *Polposipus* was reasonably straightforward to maintain and breed in captivity (with details of good husbandry methods provided by Ferguson & Pearce-Kelly 2004), and colonies have now been distributed to several other European institutions, so reducing the risk to the beetle should disease occur in any captive colony. Nearly 1000 individuals were reported to be living in captivity in early 2008. Establishment of multiple cultures of such species has many practical advantages: spreading the risk of disease or other disaster affecting any one of them, increasing the total logistic support for the species, increasing rearing space, and possibly also providing increasing potential genetic diversity by periodic outcrossings across different stocks. For some species, it might even be possible to found different colonies on different field populations.

In the mean time, rats have been eliminated from Frégate and measures instituted to prevent future reinvasion. However, because the distribution of *Polposipus* is naturally so restricted, even successful re-releases to Frégate would still maintain it as vulnerable species, and long-term conservation planning includes consideration of translocation to other islands in the Seychelles. As for any long-term captive breeding programme, useful information on basic biological features of *Polposipus* continues to be accrued.

Captive breeding is itself a far more complex exercise than commonly supposed. Many beetles have indeed been mass-reared in cages or controlled environment conditions with the aim of producing study vehicles, mass-release biological control agents, food supplies for vertebrates and other applications involving large numbers, and in smaller numbers for more specific purposes including pets, for collectors, and as public exhibits in zoos and museums (Henderson *et al.* 2008). It may be necessary to ensure specific regimes for temperature, humidity and photoperiod, together with supplies of semi-artificial diets or other foods to substitute for fully natural foods. As for Lepidoptera, much of the information on rearing particular kinds of beetles may be unpublished and in the hands of hobbyists, many of whom have substantial expertise in rearing and maintaining rare beetles of particular families, but information may also be available from studies on taxonomically or trophically related species of economic or other practical interest. However, any beetle can provide surprises! The considerable longevity of some species, whilst enhancing their attractiveness as pets or live exhibits, may be associated with generation times of several years, so that the aim of building up numbers in captivity may take considerable time to accomplish. General advice on rearing beetles, such as the survey by Singh and Moore (1985), forms an invaluable foundation preparatory to any extended captive rearing conservation exercise in which the well-being of every individual may be significant. Indeed, the details given in many of the individual species accounts by Singh and Moore span several pages and may seem excessive; they are not, and simply exemplify the considerable care and attention needed to obtain optimal outcomes.

As with insects reared as biological control agents or for other field releases, the fitness or 'quality' of captive stock must be monitored and sustained in order to counter both environmental and genetic effects in laboratory colonies. Particularly for beetles subject to unnatural conditions and foods in captivity, reduction in size, longevity, fecundity (or other performance characteristic) may occur, and may be compounded by general domestication effects and inbreeding resulting from small founder populations. A wide array of differences in environmental conditions and changes to insects resulting from laboratory rearing were tabulated by Bartlett (1985). Any such effects, as well as susceptibility to disease, may also be increased with increasing periods in captivity. Evolutionary changes in small captive populations are perhaps inevitable, but those induced in captivity may be carried into field populations on release. Any of four major categories of change may be important: (i) founder effects, with inadequate genetic variation included in the initial colony; (ii) genetic drift, whereby alleles are lost through random processes, is also of concern mainly in small colonies; (iii) inbreeding; and (iv) selection for adaptation to captive conditions. The practical advice to hobbyists provided by Walsh and Dibb (1954) remains highly pertinent at any scale of rearing: 'The whole essence of rearing [beetle] larvae is to reproduce as accurately as possible the actual conditions under which they normally exist'. However, deleterious changes do not always occur. The striking Olympic ground beetle *Chrysocarabus olympiae* has been studied extensively in captivity since the 1970s, and no progressive declines in quality were found over 13 generations, during which no additional wild individuals were imported for over a decade (Malausa & Drescher 1991).

For species of individual conservation concern, the internet can be a valuable source of information. For example, rearing protocols for many of the species of stag beetle kept as pets in Japan (see p. 120) are given in considerable detail, with much of the information broadly transferable to many other taxa. Likewise, for any phytophagous beetles, general requirements for caging and sanitation may differ little from those for Lepidoptera, and perusal of manuals on rearing butterflies and moths may be much easier than tracing more obscure and specialized literature on beetles, particularly if rapid action is needed, for example in response to finding early stages or an unexpected salvage exercise.

Threatened species of beetles may need to be maintained in quarantine facilities to help counter disease incidence and, more generally, care may be needed to (i) avoid overcrowding; (ii) provide suitable and relatively natural environmental conditions including programmed variations to parallel seasonal or diurnal photoperiod to regulate diapause and simulate conditions of proposed release sites; (iii) provide adequate, preferably natural, foods and to ensure this is available in adequate supply throughout the programme; and (iv) monitor for performance and 'quality' over generations, with contingency planning to respond to any lessening of fitness detected. Many of these measures naturally increase the costs of a rearing programme, so that careful budgeting beforehand is needed.

Releases

The availability of suitable release sites may be very limited, so that the practical exercise becomes restricted to one or a few occupied or historically occupied sites either for augmentation or introduction/reintroduction. In other cases, wider options may be available, and sites might be selected from a wider landscape on features of compatibility or correspondence with features of occupied sites. It may also be necessary to consider parameters such as altitude, because seasonal differences in development pattern may occur between lowlands and highlands (for *Carabus auronitens* in Germany, Weber & Heimbach 2001 recommended translocationg from highland to lowland sites to help compensate for this). For the rare species for which this exercise may be contemplated, some level of uncertainty over what constitutes site suitability is likely, and selection will draw largely on biological knowledge of the species' requirements, and how these are represented. A parallel context occurs in surveys for such species, where it may be very easy to overlook populations (especially if small) without very careful focus on suitable habitat. Patches may themselves be small and isolated. The remote sensing approach explored by Mawdsley (2008) (see p. 30) for tiger beetle populations, although developed for a different context, may have considerable potential for release site selection. Some guiding principles (as adopted for *Nicrophorus americanus* by Keeney & Horn 2007) can help in selecting the best option(s) for release sites.

1 Is the security, tenure or ownership of the proposed release site, and its proposed use, compatible with the beetle's future well-being and investment in its future? If there is ambiguity of tenure or use, can any suitable form of conservation agreement or covenant be devised?
2 In the case of ecologically specialized or trophically limited beetles, are there either closely related (e.g. congeneric) or trophically similar (e.g. feeding on the same plants or specific to the same prey) taxa present?
3 What levels of threat or wider disturbance may be present or anticipated and do these (e.g. weed control using herbicides) need attention prior to any introduction?
4 Are supplies of critical resources sufficient and assured? If not, is restoration or augmentation of these needed, again before introduction?
5 Are climate and topographical conditions broadly suitable, and any specific resource idiosyncrasies (e.g. bare ground for display behaviour) catered for?
6 Does the site support other notable or threatened species that enhance its conservation values and security for conserving the beetle?
7 Is the site sufficiently large to support a viable population?

In addition, the form of the release needs prior consideration: hard (direct to the field) or soft (into some field cage or enclosure to allow easier monitoring and prevent over-dispersal); single or repeated releases; and the number of individuals (also considering sex ratio and growth stage/age). A founding insect population may comprise as little as a single gravid female, but this is extreme, and considerably more individuals are recommended in order to aid genetic variability. However, for some salvage operations and translocations only a few beetles may be available and a soft release may then be needed to help prevent over-dispersal or other loss of these on the new site. For releases of *N. americanus* multiple pairs of captive-reared beetles are recommended, with an individual release consisting of an adult pair onto a corpse suitable for burial and reproduction, in some cases protecting the body from scavengers by covering with a wire mesh. This pattern reflects the beetle's normal biology, whereby a male and female bury and move the carcass together. One or both parents may then remain with the brood during development, and may protect the young from predators (Lomolino *et al.* 1995). Releases from captivity must also be planned for optimal timing. Food must be available at the time it is needed and any released stages likely to enter diapause should be able to pursue their normal phenological pattern.

Whatever precautions and care are taken, the fate of any given release is difficult to anticipate. Releases of numerous larvae of *Chrysocarabus olympiae* (see p. 144) to apparently suitable sites in the mid-1980s led to little evidence of breeding, although adult beetles were found for several years thereafter (Malausa & Drescher 1991). The 'quality' of an insect release reflects several key parameters: (i) as above and in general, the more the better, with up to several hundred individuals recommended in many insect releases for biological control establishments; (ii) genetic diversity and general health, with released individuals free of pathogens or other disease; (iii) release into climatically suitable areas; (iv) protecting the insects during transport, before release; and (v) ensuring the most suitable release method for whatever stages are liberated, and that the release site is protected adequately. Any release may or may not succeed, so that it is sound practice to retain a proportion of a captive stock as insurance and for possible further releases in the future, as recommended for *N. americanus* (Raithel 1991).

Capture, translocation or other releases involve transporting living insects, and care may be needed not to harm them by this. For example, overheating can easily occur if they are left in the sun or in a hot car for even a short time (Kozol 1991 on *N. americanus*), so that insulated holding containers may be needed. If long-distance transport is involved, for example by air or across state or international boundaries, additional measures may be necessary to avoid delay or harm: any necessary permits should be arranged well in advance; insulated containers with freezer packs may be needed; any plastic containers that might give off toxic fumes should be avoided, as should gelatine capsules that might dissolve or become adhesive when damp; care must be taken to provide sufficient food and humidity, as well as retreats to avoid cannibalism or overcrowding, and so on. For each individual exercise review of all such conditions is likely to be worthwhile and helpful in maximizing the chances of success. Releases may be timed to avoid excessive temperatures and to take place in fine weather (to avoid possible harm from rain or storms) and perhaps at particular times of the day to correspond with the beetle's normal activity pattern.

Monitoring releases to determine success or failure, and to help understand the reasons for either outcome, must be a routine activity. Again, unforeseen complications may arise. For many beetles, the presence of individuals on the release site the following season demonstrates survival to the next generation, but this is not always the case. The adult longevity of many beetles is unknown, and Weber and Heimbach (2001) summarized a number of records of European Carabidae in which adults (particularly females) are long-lived, and marked individuals have been captured after periods of 5 and 8 years. They recommended (for *Carabus auronitens*) marking individuals before release, so that such longevity can be detected in the future and separated from individuals bred on the release site. Failure of an introduction can thus be detected only after several consecutive years of absence at the release site. As another example, Balfour-Browne (1962) commented on several British water beetle species living for several years in captivity.

8

Threats or Management: the Conservation Manager's Dilemma

Ecological changes, whether natural or planned, underpin much of the need for beetle conservation, and also the core activities needed in conservation management in the field. For any exploitative land-use practice, such as forestry or agriculture, the needs for conservation are addressed most constructively by protection of some natural areas from change (as reserves) and augmenting these by increasingly well-informed and ecologically based management of production areas (Lin *et al.* 2007). In helping to conserve biodiversity whilst continuing to harvest timber from boreal forests, three complementary approaches (all exemplified earlier in this book) have been advocated (Niemela 1997).

1 Setting aside undisturbed old-forest areas to conserve specialist species and to become sources for recolonization of other, modified, areas.
2 Ecologically sound silvicultural regimes, guided by natural disturbance regimes, should be developed to augment these reserves (which are in themselves highly unlikely to be sufficiently large and representative to stand alone).
3 Restoring habitats and natural regeneration processes, using fire, to aid re-establishment of assemblages, provided that potential colonists are available.

Fire is only one of a suite of possible managment tactics for beetles. Widely used management tools, such as grazing, mowing, revegetation (or broader restoration), restriction of pesticide use and burning, all play major roles in beetle species or assemblage conservation. However, in some contexts the balance between these intended beneficial practices, particularly if undertaken without full biological understanding, can easily become harmful and enlarge the portfolio of

Beetles in Conservation, 1st edition. By T.R. New. Published 2010 by Blackwell Publishing.

threats to species coexisting with the conservation targets. Any environmental change that benefits one species may threaten others in the same place, or simply changing the regime or intensity of a management operation may change its outcome to threaten the intended beneficiary or its resource needs. For Coleoptera, by far the most intensively studied contexts are (i) Carabidae in agricultural ecosystems, stimulated largely by recognition of their value as predators of crop pests (thereby being desirable) and in turn inducing studies on their life cycles and assemblage composition as avenues towards enhancing them, and (ii) Carabidae and saproxylic beetles in relation to foresty management practices. Both have been discussed in earlier sections, with the relatively sound knowledge of their ecology providing the foundation for constructive conservation measures of much broader relevance.

A general summary of factors influencing the first of these contexts (Fig. 8.1, after Holland 2002) has wide application also to other beetles and habitats, and demonstrates the complex array of biological interactions (including those with other beetles) that determine the well-being of any focal species. From this illustration, changes in any of the categories of farming practice may be variously beneficial or harmful. From the crop-producer's point of view, the major aim may be to foster and increase populations of particular predatory species and facilitate their access to the crop. A conservationist, in addition, may seek to augment the assemblage diversity to be more representative of the presumed parental assemblage displaced by agricultural development, but there are many parallels of priority interest which converge in addressing management of agricultural areas for the benefit of beetles. Land-clearing, predominantly for

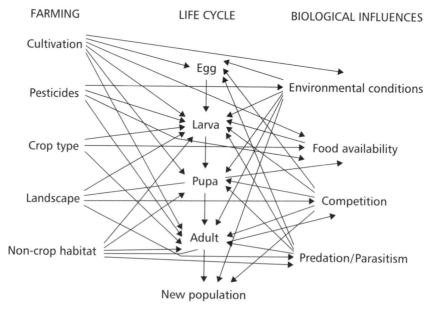

Fig. 8.1 Summary of the various influences on life cycles of ground beetles in farmed landscapes. (From Holland 2002 with permission.)

agriculture but also for a variety of less extreme transformations, has produced mosaic landscapes in many places. The extent of change differs widely. At the extremes, a predominance of agricultural systems, particularly for monoculture cropping, may leave only small island remnants of natural (or relatively natural) vegetation in a sea of modified land cover which is largely inhospitable to most native species of beetles. Towards the other extreme, stretches of forest or other native vegetation may be interrupted only by small patches of farmland. Nevertheless, the transitions of landscapes from natural to largely anthropogenic broadly follow the trajectory of dissection, perforation, fragmentation and attrition familiar to landscape ecologists. As expressed for carabids by Thomas *et al.* (2002), but equally applicable to all other beetles, this heterogeneity at all spatial scales in the landscape ultimately leads to variations in carabid reproductive, survival and mortality rates from a large number of possible influences, and furnishes the arena in which conservation studies on species and assemblages must be conducted. Management must consider scale of need.

Defining the optimal collective outcomes from management can be difficult. Thus, whereas grazing by cattle or other stock may benefit dung beetles by providing a critical resource, the accompanying changes in vegetation may threaten many other taxa that respond to vegetation condition and exposure by directly affecting vegetation composition and cover and hence wider aspects of local environments such as soil chemistry and microclimate. In this context, the implications of the intermediate disturbance hypothesis (following Connell 1978) suggest that at some intermediate level of grazing, beetle assemblages should increase in richness and diversity over ungrazed or intensively grazed areas. Low-density grazing may be associated with increased richness of Carabidae and Staphylinidae (Zahn *et al.* 2007), but specific vegetation (in this study, of abandoned wetland, reeds) may need protection from grazing in order to conserve greatest diversity. Grazing by domestic stock is obviously a condition that can be imposed or prevented in some management programmes, and the intensity, distribution and seasonality regulated. However, not only is grazing pressure difficult to quantify (usually given in number of livestock units, animal specified, per area over a given period) and variable with intensity and season, but it is also commonly not uniform across a site because of heterogeneity of vegetation structure, moisture, rocky ground and other factors contributing to fine-scale variety, so that threat versus benefit outcomes may operate on very small spatial scales in relation to the grazing imposed, to produce a very patchy habitat mosaic. Only more rarely is the level of natural grazing by naturally occurring herbivores (including insects) assessed and included in any appraisal of grazing effects.

Within reserves, imposed grazing is focused most commonly on botanical outcomes, although its importance to insects was demonstrated amply for butterflies on English chalk grasslands (Butterflies Under Threat Team 1986), where sward height was a regulator of microclimate and a determinant of habitat suitability for ant mutualists of some butterflies, and the protocols needed for grazing management to conserve particular species could be specified in considerable detail. This study clarified the differing impacts of grazing by different stock animals such as cattle, sheep and horses, a difference reported tentatively also for

Carabidae on Welsh peatlands (Holmes *et al.* 1993). However, this difference
was only tentative, as Holmes *et al.* noted, because the inferred difference between
the small effect of sheep and the much greater effect of cattle might be an
artefact flowing from estimates of grazing intensity. Nevertheless, grazing was an
important part of management of rich-fen sites for ground beetles, whilst other
members of the assemblage favoured ungrazed conditions. The major factors
influencing assemblage distribution in that study included grazing regimes (as
above), altitude, nutrient status and extent of substrate saturation. Outcomes of
particular grazing regimes are often uncertain. Comparison of extensive and intensive cattle grazing on grasslands in Hungary (Batary *et al.* 2007) suggested that
abundances of carabids, weevils and leaf beetles were not affected by grazing
intensity, although species richness was influenced in some specialist grassland
Chrysomelidae. One potential implication is that even intensively grazed grasslands may aid maintenance of rich beetle faunas, but landscape complexity in
the region may also be influential.

Most studies of grazing effects on beetles relate to cattle or sheep grazing
but, in a novel study, the effects of reindeer grazing on ground beetles and
weevils in the rather extreme environment of Finnish Lapland were studied
by Suominen *et al.* (2003). Species composition differed between grazed and
ungrazed plots on all the sites. Carabidae were more common in grazed plots
on all sites, and Curculionidae more common in ungrazed plots on birch-
dominated sites. However, in pine-dominated plots with high lichen cover (of
Cladina, a food of reindeer) the number of weevils feeding on pine was greater
in grazed plots. Overall, 18 species of Carabidae (six of them rare and not
analysed, and six found only in grazed plots with none restricted to ungrazed
plots) and nine of Curculionidae (five of them rare and five only in grazed plots
with again none only in ungrazed plots) were found. Reindeer grazing had
strong impacts on the ground-dwelling members of these families, and even fairly
high grazing pressure may increase diversity and richness of their assemblages.
Suominen *et al.* (2003) suggested that the effects of reindeer grazing were probably indirect, through foraging and trampling leading to changes in vegetation
and the physical character of the sites.

Trampling effects on exposed riverine sediment (ERS) beetles were noted
earlier (see p. 87) but, in general, direct trampling effects on ground-dwelling
beetles have only rarely been quantified. One anomalous and informative result
reported by Sadler *et al.* (2004) was that intensity of livestock trampling on
ERS beetle habitat was associated positively with the number of ERS beetles
of conservation interest. The tentative explanation for this was that trampling
destabilizes the substrate, suppressing vegetation succession and so rendering more
space available for beetles as a positive outcome along rivers that otherwise have
little suitable habitat available. Increased species richness with higher trampling
rates (of sheep and cattle, based on livestock unit counts and faecal counts as a
second index) was related also to increased levels of organic matter and silting.
At least some of the associated beetles were linked biologically with these features. However, the ERS quality score (see p. 45) was considered a better index
of conservation value than species richness, because it explicitly considers the
rarity of each specialist beetle species present. On that basis, trampling indeed

reduced the conservation values of ERS beetle communities, endorsing need for restriction of this activity.

Measures to conserve beetles usually involve, or necessitate, modifications to a local biotope or wider landscape, depending on the scale (single species to assemblages, local to regional or national need) and purpose (maintain or increase local carrying capacity, increase size and number of populations, found new populations, etc.) of the exercise. It is almost inevitable that ambiguities and conflicts of purpose may arise, perhaps particularly when the areas selected or needed for conservation management are also desired for industrial purposes. In such context, the presence (or even just the suggested or implied presence) of species (including beetles) declared formally of conservation concern can conflict with commercial forestry, mining or farming, or with commercial operations to control important pests (also including beetles) in which absence of non-target effects cannot be guaranteed. Any such case necessitates conservation measures incorporating, and orchestrated with, measures to mitigate any adverse impacts, in a broad management agenda. Thus, the Tasmanian stag beetle *Lissotes menalcus* (Mt. Mangana stag beetle) is entirely dependent on a supply of dead logs for breeding sites, and any land-use activity in forests that either affects the existence of this resource or interrupts the future supply of decaying logs will threaten local populations (Meggs & Taylor 1999; Meggs 2002). The drive to convert native eucalypt forests to pine plantations for softwood production in Tasmania (see p. 186) may also lead to local extinctions of Simsons stag beetle (*Hoplogonus simsoni*) (see p. 181), so that conservation recommendations include limiting the impact of pine plantations on the beetle but involving prescription forestry rather than total prohibition of plantations. More generally, provision of resources for saproxylic beetles in areas used for forestry may influence cutting patterns and the disposal of coarse woody debris to maintain a mosaic habitat, two examples being the need to leave both standing and fallen dead oak trees, as the trees in different situations support considerably different assemblages (Franc 2007); and leaving high stumps when felling (cut at 3–5 m, leaving 3/ha of clearcut forest) (Schroeder *et al.* 2006). Dead oakwood is regarded as one of the most important resources for saproxylic beetles over much of Europe, and Franc's study strongly endorsed the need for fallen logs to be retained or continuously created in managed forests. High stumps (usually of conifers) are also desirable, and the presence of stands containing both forms of dead-wood habitat from several tree species is likely to prove particularly valuable for conservation.

Any forestry practice, such as clearfelling, has complex ecological effects and it is perhaps too simplistic to relate, for example, numbers of carabid species solely with cleared areas for conservation evaluation. Several studies (including those by Butterfield *et al.* 1995 and du Bus de Warnaffe & Lebrun 2004; see p. 92) have shown increased richness in cleared areas with many more rare species in clearfellings than in mature stands (Butterfield *et al.*). Clearcut areas are colonized by open-habitat species, and are highly important for conservation of many beetles and other invertebrate groups. However, the creation of clearcut areas goes hand in hand with major disturbance to the parental forest, with likely loss of beetles that depend on closed forest environments. Old-growth specialist

Table 8.1 Percentage occurrence in each regeneration age class of the five most abundant carabid species in the Picton Valley forest, Tasmania, Australia.

	Regeneration age class (years)			
	Early (1–3)	*Inter (7–10)*	*Late (20+)*	*Old growth*
Sloaneana tasmaniae	10	70	10	10
Rhabdotus reflexus	10	60	20	20
Promecoderus spp.	2	80	7	12
Mecyclothorax ambiguus	90	10	0	0
Notonomus politulus	26	7	1	65

Source: Michaels & McQuillan (1995) with permission.

species are particularly at risk. In addition, clearcut areas are often only temporary open habitats (for a maximum of 10–15 years; du Bus de Warnaffe & Lebrun 2004), so that resident species will then need to move, and ensuring that such sites cannot replace more natural open areas such as meadows or succession-limited grasslands. Their value in forested areas depends on continuity, perhaps through design of a successional mosaic. Open-habitat carabids are often rare in forests. However, large clearcut areas in forests may also help to conserve some species from surrounding open habitats if these become threatened by urbanization or intensive agriculture. The large carabid *Carabus cancellatus* is one such beneficiary in Belgium (du Bus de Warnaffe & Lebrun 2004).

The converse to the requirement for mosaics of open areas in forests is the complementary need for areas of mature remnant forest to be retained, as well as a successional mosaic of different cutting stages. Comparison of Carabidae across four stages of cutting in *Eucalyptus obliqua* forest in Tasmania (Michaels & McQuillan 1995) revealed the five most common species to be represented very unevenly across age categories, with 60% or more of each species trapped in one successional stage (Table 8.1). The need for mature remnants is encapsulated in formal practices such as use of wildlife habitat strips. These are corridors of remnant forest retained among plantation forests, in Tasmania at least 100 m wide and left every few kilometres throughout the production forest, whether this is of native or introduced tree species. In some plantations these may be the only remaining areas of native forest. Their values for beetles were investigated by Grove and Yaxley (2005), who used pitfall traps in plantations, wildlife habitat strips and native forest to demonstrate considerable differences in assemblages. Assemblages in the strips resembled those near the edge of mature forest more than those in its interior, so that they may not be effective for interior-forest beetle conservation, but still support many species rare or absent from plantations, with the strips harbouring assemblages intermediate in composition between those of plantations and native forest. Grove and Yaxley suggested that, at least for the wet sclerophyll forests of Tasmania, wildlife habitat strips should be augmented by larger reserves and networks of continuous native forest dispersed throughout the production forest landscape in order to augment the interior habitat need for specialist forest beetles.

Wildlife habitat strips are simply one example of incorporating conservation needs into formal guidelines or acts that determine or influence forestry practices in many parts of the world. Practices for conservation of native species are integral in many acts. For example, the Swedish Forestry Act obliges or strongly encourages management such as creating snags, saving trees along banks of waterways, or leaving standing or fallen dead trees in place (Johansson *et al*. 2006). The counter is that the same Act permits foresters to leave only 5 m^3 of fresh conifer wood per hectare per year, this limit being to restrict build-up of pest Scolytidae. Johansson *et al*.'s study implied that this restriction could be abandoned in northern Sweden, where cool climates naturally restrict scolytid breeding, a situation that might however change as climates warm in the future.

Fire

The effects of prescribed burning and, more generally, use of fire in management of vegetation constitute one of the most difficult areas of conservation ecology upon which to generalize. These effects are undoubtedly complex, and most authors have been unable to assess all the key variables of frequency, area, intensity, intervals between fires and seasonality at the same time. Thus, Moretti and his colleagues in Switzerland examined the richness of several beetle families in unburnt sites (with no fires for at least 35 years), sites with one fire over that period and sites with repeated (three to four) fires (Moretti & Barbalat 2004; Moretti *et al*. 2004). This correlative study suggested that fire might have positive effects in increasing richness of Carabidae, with a trend towards an increase in saproxylic beetles (Cerambycidae, Buprestidae, Lucanidae). Conversely, fire had a negative effect on weevils, probably through decreasing vegetation diversity. The literature on fire effects on insects is both voluminous and confusing, with inferences or outcomes reflecting some or all of the above parameters of a fire regime, together with the vegetation type and how outcomes are monitored and interpreted. Scale of burning is of central importance, in that mosaic burning will create an array of intermingled successional stages/habitats that by definition will not be found in large uniformly burned areas. For example, prescribed regimes for burning of clearcut forest areas as preparation for planting or seeding vary substantially with context and region. Some workers have compared incidence and abundance of beetles (mostly Carabidae) and may include additional treatment effects in field investigations. In Norway, Gongalsky *et al*. (2006) compared burned and unburned forests, with additional treatments including partially clearcut, selectively cut and entire forest; this revealed that overall richness (with a pool of 32 ground beetle species) was least in unburned forest, with progressively greater numbers in burned standing forest, burned selectively cut forest and burned clearcut, which had the most species. In this example, inferences included that the burning of single stands may have conservation benefits, whilst larger-scale fires might be even more useful. There is little doubt, however, that fires can be very destructive to beetles, with the importance of refuges impossible to overstate: beetles that are undergound durng a fire, for example, face a very different scenario from those exposed on flammable low vegetation.

Forest burning may kill trees or otherwise condition them to attack by a variety of saproxylic beetles, with some taxa specialized to respond and seek out recently burned trees (see p. 189). Low-intensity fire is used extensively in forests as an important management tool to reduce risks of higher-intensity wildfires by reducing fuel loads on the forest floor and, in general, such managed fires are undertaken only every decade or so, although such prescriptions vary widely. Controlled understorey burning, as undertaken in boreal forests in western North America (Niwa & Peck 2002), can reduce fuel loads effectively, but may also greatly modify the habitat for understorey and litter invertebrates. Many studies on the purported effects of fire on insects are based not on temporal before-and-after surveys on the same sites but on comparisons between unburned (control) and burned sites over the same (post-fire) period, with the assumption that these sites earlier supported the same (or a closely similar) suite of taxa. Reflecting the substantial spatial heterogeneity of beetles and other insects, this assumption is sometimes difficult to accept uncritically. Nevertheless, with sufficient replication some consistent trends may emerge. In a comparison of nine paired burned/unburned sites in Oregon, collectively yielding 14,793 Carabidae of 13 species (with another four represented by singletons), the mean catch in unburned sites was about twice that of burned sites, but no species disappeared entirely from burned sites, and one (*Trachypachus holmbergi*) occurred, albeit uncommonly, only on burned sites. Species richness and diversity indices were similar across treatments (Niwa & Peck 2002). The trend of reduced carabid numbers in burned (by wildfire) forests was also found by Holliday (1992), and it is likely that wildfires would generally be more intense than the control burns of the Oregon study. Several of the most abundant carabids are opportunistic foragers, so would not necessarily be affected by fire destroying any specific food materials. Likewise, most of those species were believed to be spring breeders so that timing of control burns, at least on some sites, may not eliminate them.

The edges of fires in forests are likely to be important ecotones, so that several workers have suggested that such irregular edges should be emulated in landscape management. The structure of wildfire/forest boundaries in relation to those caused by mountain pine beetle (see p. 3) led McIntire and Fortin (2006) to advocate use of irregular edges in forest management for biodiversity conservation. If fires occur at sufficiently small scales to help create and maintain mosaics of forest patches, as implied for winter fires in the Swiss Alps (Moretti & Barbalat 2004), they may help to foster beetle species richness, particularly by opening up small areas of vegetation. Fires in forests are important in continually providing ephemeral habitats such as exposed soils and freshly killed trees for saproxylic beetles and, in conjunction with protection of remnant patches from any burning, may help to encourage the landscape diversity essential to sustaining high diversity. A general hypothesis applicable to many beetles is that locally high populations developing after fire allow species to persist in suboptimal unburned habitats (Jonsell *et al*. 1998). Nevertheless, recently burned boreal forest patches are key habitats in succession (Saint-German *et al*. 2004, for

Canada), where fire-associated Coleoptera include several genera of wood-boring Cerambycidae, and stressed-root-seeking Elateridae. The assemblages differed substantially between burned and unburned stands in that survey, but many gaps in understanding their dynamics persist. Regimes of prescribed burning in forests may be routinely accompanied by other landscape changes, such as maintenance of firebreaks; thus the tiger beetle *Cicindela patruela consentanea* in the eastern USA benefits from prescribed burns providing open habitats, as well as from routine maintenance of firebreaks and access trails (Mawdsley 2007).

In contrast to much management burning of forests, burning of grasslands or savannas may be much more frequent. Thus, half or more of the vast savanna landscapes of northern Australia are burned annually (Orgeas & Andersen 2001). Reconstructed prairie grassland in Iowa, USA supported lower diversity and evenness of carabid assemblages in plots burned a few months previously than in plots burned 2 years before sampling (Larsen & Williams 1999), although species richnesss was greatest in the most recently burned plots, again reflecting presence of opportunistic species. In general, little is known of the effects of savanna fires on invertebrates, with the variety of studies undertaken inferring an equivalent variety of insect responses and changes. One important and informative trial, undertaken at Kapalga (Kakadu National Park in the Northern Territory of Australia), indicated some of the complications involved in interpreting these. Replicated fire regimes on large (15–20 km^2) landscape units were assessed under three treatments: (i) annual burns early in the dry season (May/June); (ii) annual burns towards the end of the dry season (September/ October); and (iii) unburned, with fire excluded. The background to this long-tem experiment was given by Andersen *et al.* (1998). Sweep-net sampling of beetles yielded 233 species across 26 families, with the sample of 3865 individual beetles dominated by Chrysomelidae (91 species) and Curculionidae (41 species), these collectively comprising 91% of individuals. Despite some differences across treatments, the assemblages appeared to be very resilient to fire, probably reflecting a long history of association with frequent burning. The importance of fire appeared to be secondary to that of variations in rainfall and soil type. Changes in assemblages were evident only after repeated fires, so that it is necessary to consider cumulative effects rather than simply those of any particular fire, as is most common in studies of this nature. The Kapalga experiment contrasted the extremes of no burning and annual burning, and only one of the 10 most common beetle species was affected significantly by any treatment.

Pitfall trapping was also used at Kapalga, over 7 years, and beetle abundance in dry-season sampling (July–August) was too low for any clear analyses. Wet-season trapping showed fires late in the dry season to affect the richness of ground-active beetles, but this was influenced also by rainfall prior to sampling, so that an important message to managers is that rainfall must be included in the sampling design in order to aid interpretation of fire impacts (Blanche *et al.* 2001). The pool of surface-active beetles at Kapalga was 200 species or morphospecies across 39 families, with the richest being Staphylinidae (30 species), Carabidae (29) and Scarabaeidae (26). Burning late in the dry season has the potential to shift assemblage composition to contain an increased proportion of dry-adapted beetles.

Manipulating beetle populations

Much of the applied ecology of beetle species has been directed towards a wish to change their populations, either to foster and conserve beneficial species or to suppress pests, and both native and alien species are involved in either category of venture. The relevance to conservation exercises primarily stems from much of the above effort being focused progressively on ecological understanding and environmental safety. The beneficial species are almost all predators, more rarely herbivores, and encompass alien species introduced as classical biological control agents and native species encouraged in the context of conservation biological control. Pest beetles, in contrast, include representatives of most trophic groups but are dominated by herbivores in their various guises. Lessons from the many studies on these taxa provide much information of direct value in conservation, both in directing what should be done and what should not be done in manipulating beetles of conservation interest and demonstrating the variety of approaches that may be available or suitable for their management or recovery.

The dilemma, and need for compromise, is well illustrated by some aspects of management in boreal forests. The balance is between thwarting pest beetle attack and conserving the numerous other saproxylic beetles by removing or providing dead wood in the forests, but the solution raises complex issues. The spruce bark beetle *Ips typographus* kills vast numbers of trees and is responsible for providing a high proportion of dead wood in some forests: one published figure, for reserves in Sweden, is 81% of standing trees (Schroeder 2007), with the remainder mostly comprising storm-felled trees. Overall, the amount of coarse woody debris in Swedish forests has been reduced by about 90% through forest management compared with levels in unmanaged stands. Salvage logging of dead trees is undertaken for several reasons: (i) rescue of economically valuable timber; (ii) reducing the risk of bark beetle attack in nearby forest stands as beetles emerge; (iii) reducing the amounts of fuel and so reducing intensity and risks of forest fires; and (iv) accelerating forest regeneration. Fungus transmission by *Ips* (see p. 130) can reduce the value of beetle-killed trees markedly, and Schroeder advocated retaining killed trees in many situations as a cost-effective way to augment the supply of dead wood. Moreover, because many salvaged trees are not removed before much or all of a new generation of beetles has emerged from them, the aim of preventing beetle infestations may not be fulfilled, and the trees could be left in the forest without them then contributing to further beetle attack. Trees salvaged in this way are also reduced in value if the bark falls off the trunks (e.g. through increased woodpecker activity; Fayt *et al.* 2005) during logging, so that the beetles are not then actually removed from the forest; again the trees could be left on site (Schroeder 2007).

This dilemma has also been addressed in a related context, that of piles of harvested oakwood left in forests before being chipped and used as biofuel (Hedin *et al.* 2008). The piles contain tops, branches and small trees with branches and twigs, so that a considerable variety of wood textures and diameters are accumulated in piles that may be up to about 65 m long and several metres in height and depth. Saproxylic beetles were reared from representative wood samples from piles at 12 sites in southern Sweden, yielding 39 species, including several red-listed species. The great majority of individuals (77% of the total of 3528) comprised the bark beetle *Scolytus intricatus*. Hedin *et al.* suggested that these fuel piles are ecological traps for saproxylic beetles (as sink habitats) if they are chipped before the beetles emerge. Management considerations include recommendations to pile the wood outside the beetles' major adult activity seasons, namely winter or early spring. In such cases, retention of the top 10% of the pile may also retain many of the beetles in the area, whereas with longer infestation periods or with cut wood exposed earlier before being stacked, this option has less benefit. Suggestions of longer storage, so that primary colonizer beetles have developed fully, do not account for concerns over later-arriving secondary species. In stands with known conservation values, for example by harbouring red-listed species, wood from cuttings should preferably not be removed for fuel.

Removal of logging residues for biofuel also influences ground-active beetles. In a comparison of Carabidae in clearcut spruce plots in central Sweden, either with all debris left on site or with the slashed material removed, the richness and diversity were greater on cleared sites, with a greater proportion of generalist species and reduction in forest specialists (Nitterus *et al.* 2007). The effects of slash removal may persist, with effects found 5–7 years after harvest, so that the conservation of specialist species may be aided by leaving the material in place: Nitterus *et al.* suggested leaving it on every fifth clearcut plot. From other studies (such as that by Koivula & Niemela 2003) it is well known that the amount of logging residue left in forest affects the structure of carabid assemblages by augmenting habitat complexity and variety.

More generally, some forms of disturbance to forests and other biomes are natural, with many resident species (including saproxylic beetles) adapted to the varying supply of key resources engenderd by disturbance events. The major pressures incurred by recent human activity thereby relate to unusual disturbance, either in intensity, frequency or novelty, so that clearfelling of large areas of forest is not a normal occurrence. As Kaila *et al.* (1997) noted for boreal forest beetles 'natural disturbances, such as forest fires, storm damage . . . create open, sun-exposed areas with a considerable amount of dying and dead wood . . . regionally they may always have been present with moderately small interpatch distances, easily covered by the dispersal capacity of saproxylic beetles'. Such natural disturbances have now been largely eliminated from boreal forests by larger-scale anthropogenic changes. A general and widespread conservation requirement is that forestry practices should emulate natural processe as closely as possible. For boreal saproxylic beetles, a key aspect appears to be availability of decaying wood in open conditions, with some beetles depending on such sun-exposed habitats. The red-listed *Atomaria subangulata* (Cryptophagidae) is one such beneficiary (Gibb *et al.* 2006).

Other management trends likely to be beneficial include provision of coarse woody debris in clearcut areas in plantation forests, as noted above, and strategic planting of trees to create varying ages and reduce stand-age uniformity. As Davies *et al.* (2008) illustrated, some threatened beetles may be able to withstand clearfelling as long as their dead-wood resources remain available. Progressive conversion of older uniform-aged plantations into more varied forest mosaics is receiving greater attention as awareness of the conservation values of variety increase. Nevertheless, for any particular beetle species, the detailed information needed for a robust conservation prescription is usually not available (Davies *et al.* 2008) so that management is necessarily somewhat generalized. Clearcut forestry has been criticized for failing to mimic natural processes adequately, so that moves towards reducing this are increasing. Patch retention of natural or older forests in plantations has been advocated widely, with some studies of Carabidae in such retained patches casting doubts on categorizations of some beetles as forest specialists (see Lemieux & Lindgren 2004). Saving trees from clearfelling for conservation in boreal forests extends well beyond sparing old trees. Aspen (*Populus tremula*) is often retained and provides substantial values for many species in helping to assure future supplies of coarse woody debris, with increased aspen recruitment (from improved forest management and fire suppression) an important facet of habitat maintenance for saproxylic organisms (Sahlin & Ranius 2009).

Habitat restoration

Lessons from ecological manipulations, such as the influence of beetle banks and crop margins (see p. 173), have demonstrated approaches of far wider import-ance in re-creating habitats for beetles in the landscape. For example, relatively simple measures, such as establishment of hedges in agricultural landscapes, can produce rapid and substantial effects on carabid diversity by providing easily colonized habitats suitable for many species not adapted to more open field habitats (Fournier & Loreau 1999). Four groups of ground beetles were distinguished in that survey, namely species restricted to the hedge, species pre-ferring (i.e. more numerous in) the hedge, species preferring the adjacent crop, and species unaffected by the hedge. Structure and composition of vegetation may be very important, as well as facilitating the well-being of many taxa beyond beetles, and improvement of vegetation condition is a primary aim in the con-servation programmes for many terrestrial beetles, and often also relevant for aquatic species. Richness of beetles in a biotope may sometimes be related to plant species richness (Buse 1988), with groups of beetle taxa related to all kinds of vegetation associations, not necessarily to single plant species. Restoration or rehabilitation of landscapes, or even of small defined sites, can be a lengthy process, and in practice the effort may need long-term commitment, for example to modify normal patterns of succession rather than to more simply enhance supplies of single food species, in itself often a complex task. Details of any such programme differ with the taxa and the condition of the site, as well as being dictated by the conservation objectives, which should be clearly defined and communicated. The spectrum of activities ranges from simple enhance-ment of critical resources for individual species to wider restoration of sites to accommodate typical assemblages. The purpose might also vary and dictate the priority actions: preparing a site for an introduction of a species and for natural colonization of numerous species may demand somewhat different approaches and monitoring.

The significance of major landscape features (corridors, nodes, gaps, etc.) and vegetation features (woody or cleared, densely vegetated) on particular beetles affords an opportunity for planning based on experimental studies of the beetle's responses, as indicated earlier. Simulation studies of behaviour (based on real measurements) can be informative. Individual behaviour of the carabid *Abax parallelopepidus* in a landscape was discussed by Burel *et al.* (2000, drawing on earlier work, including Charrier *et al.* 1997), involving eggs, sub-adult and adult beetles, located on a landscape map where hedgerows, woodlots, pastures and crops were identified. Effects of landscape structure were then appraised by simulations of models incorporating changes in parameters such as density of hedgerows and quality of hedgerow vegetation. Hedgerow structure is an important determinant of corridor function, and a dense herb layer and presence of a tree layer are both beneficial features (Burel 1989). Beyond about 1 km from a forest edge, the presence of forest carabids depends largely on such features, and their well-being is enhanced further if two hedgerows abut a laneway.

The universal requirement for any landscape management scheme is that of defining optimal landscape mosaics to sustain natural biodiversity in the context of also satisfying human needs. Three major tools are available for this, namely experimental assessment (to which many studies of beetles have contributed), inferences based on all (or at least the best available) ecological information, and modelling (again, based on realistic parameters or real data). The subsequent challenge of management involves using spatial heterogeneity as a benefit in conservation and, as Banks (2000) implied, this requires moving beyond the idiosyncrasies of individual species responses to more general theory that could be implemented in a variety of agricultural (and other) settings and which achieves an optimal compromise between the various competing interests involved. Thus, much restoration of landscapes involves trees, either purely as natural features or as components of afforestation as investment in future industry. Conifer afforestion in northern England can be designed in various ways (Buse & Good 1993), to incorporate different plantation sizes and shapes, tree species, age classes, use of rides and roads, collectively buffering or increasing connectivity of existing tree patches with the functions of increasing amount and quality of habitat for beetles. In Buse and Good's example, many species of Staphylinidae may benefit from enhanced habitat diversity, which can result from design considerations without compromising the major commercial purpose of the exercise or increasing costs unduly. Some staphylinids, for example, benefit from forest-edge habitats, so that small habitat units within conifer plantations could be a positive conservation step. Mosaics of age and condition have wide benefits for many beetles, as many of the studies cited earlier in this book have shown.

Reforestation is used widely in commercial forests or as a conservation measure to replace forest cover after its loss. In eastern Australia, ground-active beetle assemblages were compared across five different reforestation styles, and within cleared pasture and intact rain forest. All the assemblages in reforested

Table 8.2 Representation (as percentage distribution) of the more common species of Carabidae (those with 4% or greater part of the catch in any plantation age group) in soil samples from plantations of *Picea sitchensis* in Britain at three different stages of the forestry cycle.

	Plantation age (years)		
	0.5–4	*5–22*	*42–63*
Trechus spp.	74	32	78
Notiophilus biguttatus	2	+	14
Bradycellus harpalinus	4	11	1
Amara apricaria	0	13	0
Pterostichus diligens	2	5	0
P. strenuus	+	6	0
Trichocellus cognatus	1	6	0
Bradycellus ruficollis	0	5	0
No. of individuals	206	289	91
No. of identified species	15	22	5

+, presence in catch at < 1%.
Source: after Butterfield (1997) with permission.

sites were intermediate between those of the two extremes (Grimbacher *et al.* 2007), with structural complexity in the assemblages progressively approaching that of intact rain forest as the complexity of reforested sites increased. An important implication from this study is that high-quality habitat for rain forest-dependent beetles is difficult to replace from timber production forests, in part because harvest cycles may be too short to achieve canopy closure and the conditions that the original beetle assemblages may need. Even complex restoration plantings may not be an effective short-term or medium-term substitute for retention of intact rain forest. More generally, the objectives of restoration, and the trajectories needed in conservation, may be designed in part from studies of assemblages or key species in degraded and pristine habitats, particularly if data tracing changes over a forestry or other perennial cropping cycle are available. Carabid assemblages in Britain may differ substantially between clearfell and conifer plantations (Butterfield *et al.* 1995) along the successional change in *Picea sitchensis* plantations within a normal forestry cycle (discussed also by Butterfield 1997). She compared ground beetles in recently clearfelled areas (within 1–4 years of clearing as preparation for second-rotation *Picea* planting), young (5–22 years after planting) and old (42–63 years after planting). Major changes in assemblage composition occurred over this cycle, as indicated in Table 8.2, but ordination analysis revealed the likelihood of rapid changes in earlier stages and greater stability in older stands, in which the number of species was considerably less. With canopy closure in old plantations, the herb layer is reduced and ground-level temperatures may be buffered, influencing which beetles can persist. The extent of exposure also influences many saproxylic beetles (see p. 84), as sun-exposed dead wood is attractive to many species and is thus an important resource to conserve, whether generated by fire or felling. In trials involving four tree species in northern forests, 316 beetle species (including 40 red-listed species) were captured as they emerged from net-enclosed stumps (Lindhe *et al.* 2005). Two-thirds of the 86 more frequent species were more commonly found in exposed (clearfelled) or semi-exposed (thinned) plots, whereas only one-third were more frequent in shade. Augmentation of exposed dead wood from forestry practices might thus be an important conservation measure, both to provide habitat and to increase continuity with retained forest. The other important implication was that many beetle species in the forests might benefit from prescribed burning, and from maintaining an open forest structure (Lindhe *et al.* 2005). Although relatively few beetles favoured shaded wood, this resource is still important to conserve. Some Scolytidae prefer moist and shaded habitats, with artificially shaded logs sometimes more attractive than those in natural shade, as darker and moister environments (Johansson *et al.* 2006).

Similar principles apply to aquatic beetle environments, where some guiding information may be available from knowledge of the insects, but much of the process may involve original design. In Irish peat bogs, vegetation structure and diversity appeared to be key features in sustaining high richness of water beetles, but Cooper *et al.* (2005) noted that data available were insufficient to determine management practices for individual beetle species. Broader management prescriptions emphasized the importance of heterogeneity, as noted above for terrestrial biomes, with suggestions to 'reduce variation in the structure and spatial variation of drain and peat-pit habitats by peat-pit excavation and drain clearing', whilst maintaining present pits separated from agricultural drains to help maintain present nutrient levels (Cooper *et al.* 2005). Restoration of aquatic and riparian habitats for beetles has attracted considerable attention, perhaps particularly for riparian Carabidae, to counter the profound changes that have occurred in their natural habitats through changes in watercourse dynamics. Monitoring beetles after flood-plain restoration practices in Germany (Gunther & Assmann 2005) showed that many stenotopic carabids could colonize new habitats rapidly, in some cases within a few weeks or months, leading to establishment of breeding populations there. Changing inundation regimes may be important for some ground beetles. Some species can withstand up to several months of inundation (Fuellhaas 2000), and may be useful monitors of changing water management regimes designed to promote heterogeneity through restoration. However, varied responses occur to inundation, with dry-meadow species responding negatively and hygrophilous species positively on the fen grassland studied by Fuellhaas. Several riparian carabids of conservation interest in Europe may need to be reintroduced to such habitats after they have been restored, because the beetles disperse only weakly and may not be able to naturally reach the areas they inhabited in historical times. Inundation may be a severe threat to both riparian and true water beetles, as Turner (2007) showed for the impacts of reservoirs on Table Mountain, South Africa. Many of the beetles there are local endemics and of conservation importance, and species loss was attributed in part to change from a riverine to a lacustrine habitat, with other factors (water chemistry, temperature, hydrology) also probable influences. Habitat restoration

Table 8.3 Anthropogenic and natural habitats for Carabidae: examples of artificial analogues of some natural habitats.

Natural habitat	Anthropogenic analogues
Saltmarsh	Flooded colliery spoil Salt-pans
Fenland	Wetlands on pulverized fuel ash from power stations Flooded sand quarries
Calcareous grassland	Lagoons of dried river dredgings Dry colliery spoil Lime kiln waste Leblanc process waste
Heathland	Abandoned sand or gravel pits Perpetually disturbed road verges on sandy soil
Inland cliffs and scree	Hard-rock quarries Demolition sites Industrial installations Railways
Open woodland	Hedgerows
Caves, mammal burrows	Cellars, stables

Source: Eversham *et al.* (1996) with permission.

or offsets may not be adequately served by approximations, but any restoration project that extends beyond framework habitats may become too expensive and difficult to pursue.

The large scale of human manipulations to natural habitats ensures that much of the area at present available for beetles is to some extent artificial; indeed, in places such as Britain, truly natural habitats are scarce or, according to some opinions, even non-existent. Conservation must then devolve on anthropogenic habitats, and Eversham *et al.* (1996) emphasized the value of anthropogenic analogues of natural habitats and that they can indeed support rare and threatened beetles. The categories of these anthropogenic analogues are exemplified in Table 8.3, many of them paralleling brownfield sites (see p. 107) in being post-industrial areas of quite recent origin. They range from wetlands to dryer biotopes, and were collectively differentiated from intensive agriculture/forestry areas in that the soil type is either artificial (such as pulverized fuel ash) or would not occur without human activity or industrial inputs. Eversham *et al.* reported that more than 35% of Britain's rare and scarce carabids have been recorded on such areas, with some assemblages notable for their restricted distribution and content of significant species. Many potential sites are largely derelict at present, and their incorporation into the conservation estate might involve only limited restoration efforts and help to maintain the early successional stages that have been passed by in many reserves (Eversham *et al.* 1996).

A deliberate policy of mitigation, deliberately replacing resources lost to development, is a strongly focused and controlled form of habitat restoration, and is included in conservation and management for the cerambycid *Desmocerus californicus dimorphus* (Valley elderberry longhorn) in California (see p. 128). Although other factors also contributed to decline, a major management action has involved extensive replacement planting of the beetle's host plant (*Sambucus mexicanus*) on riparian sites (Collinge *et al*. 2001; Holyoak & Koch-Munz 2008). Initially many of the planted sites were not colonized, so that original guidelines devised in 1988 have been modified progressively to increase planting densities, increasing from two to 10 new elderberry stems for every stem lost, and also including planting of associated native shrubs and trees. Many of these mitigation sites are much smaller than natural riparian patches occupied by the beetle, but a range of factors influences beetle density. In general, larger mitigation sites supported more beetles than smaller mitigation sites, so that management lessons include not to unduly fragment natural riparian habitat in the future (Holyoak & Koch-Munz 2008).

9

Conservation Lessons from Beetles

Several groups of beetles have dominated much of the earlier chapters of this book and, in contrast, many families of beetles have been omitted completely or received only very scant mention. The dominant groups are those that have received most ecological attention or been favoured by hobbyists so that accumulated information on distribution, biology and, where necessary, conservation need can be appraised with some confidence. Some further general perspective on a number of these predominant groups is given in this chapter. Available information on many other beetles is much less comprehensive and whilst biological and distibutional frameworks for many have become sufficient to indicate the rarity or vulnerability of many northern temperate region species in particular, many of these have not contributed substantially to the emerging wider picture of beetle conservation need. Some such lacunae are unexpected, for example the vast diversity of weevils and leaf beetles, both of which include substantial numbers of pest species that have individually received considerable and sustained biological attention aimed at managing them constructively, is not reflected in the conservation literature.

Water beetles

Most groups of beetles selected for individual attention here are natural taxonomic entities. Water beetles are not. They comprise an array of families, spread across both major suborders of Coleoptera, and united by their dependence on aquatic habitats, their appeal to the same group of enthusiasts, and the conservation problems and concerns attendant on the well-being of freshwater bodies and wetlands. The very term 'water beetles' becomes somewhat subjective,

Beetles in Conservation, 1st edition. By T.R. New. Published 2010 by Blackwell Publishing.

depending on the individual worker's inclination to include, for example, species characteristic of ecotones between freshwater bodies and adjacent land, but not necessarily with any aquatic life stage. Thus riverine Staphylinidae and Carabidae were included as water beetles by Ribera (2000), but two families of Byrrhoidea (Limnichidae, Heteroceridae) were not. Several large beetle families (Staphylinidae, Carabidae, as above, Chrysomelidae, Curculionidae) are almost wholly terrestrial, but contain species associated with aquatic habitats. The taxonomic scope of this group is thus rather variable, and distinct from the majority of terrestrial beetles. Beetles have colonized the aquatic environment on several independent occasions, so that (as noted above) different major phylo-genetic lineages occur there. The five major lineages, as listed by Ribera *et al.* (2003), are the Hydradephaga (families Gyrinidae, Haliplidae, Noteridae, Hygro-biidae, Dytiscidae), Hydrophiloidea (Helophoridae, Spercheidae, Hydrochidae, Georissidae, Hydrophilidae), Dryopoidea (Dryopidae, Elmidae), Hydraenidae, and Myxophaga (Microsporidae, Hydroscaphidae). Nevertheless, as Foster (2000, p. 224) put it, 'A precise count for water beetles is impossible simply because a water beetle cannot be precisely defined'.

Several of the families conventionally included in this artificial array are small, but others are very widely distributed. For the Iberian Peninsula, Ribera (2000) included members of 20 families (totalling 100 genera) with the more than 600 species included within a wider Mediterranean fauna suspected to be around 1000 species. The most diverse families in the Iberian Peninsula are Dytiscidae (estimated by Ribera as 33 genera, 163–165 species and subspecies), Hydraenidae (6 genera, 138 species), Hydrophilidae (20 genera, 90–92 species), Helophoridae (1 genus, 35 species) and Elmidae (10 genera, 29–30 species). Although several groups of primarily aquatic insects (particularly Ephemeroptera, Plecoptera and Trichoptera) are assessed widely in evaluating the condition of freshwater ecosystems, Eyre *et al.* (2003c) noted several advantages of water beetles for environmental monitoring (including that 'they tend to be the only macroinvertebrates present in most freshwater habitats') and suggested that they could be accorded freshwater values parallel to those of Carabidae in land-use planning for terrestrial habitats. In the Segura river region of south-eastern Spain, Coleoptera were richer than any other invertebrate group sampled by Sanchez-Fernandez *et al.* (2006), with 147 species in 52 genera. The next most abundant group, Trichoptera, contained 51 species in 32 genera, and beetles were collected in all 40 sites investigated. Trichoptera, in contrast, occurred in only 21 sites, so that both richness and ubiquity of aquatic Coleoptera suggested their possible values in conservation assessment in this region. They were the 'most reliable' surrogate group for wider biodiversity in this study, by correlation with a remaining richness value (calculated as the total number of species of all focal groups assessed at a site minus the number of species of the focal group, here beetles).

Part of the value of water beetle assemblages as putative indicators arises from the broad array of taxa and feeding habits represented. Typical assemblages are likely to include a spectrum of predators, herbivores, algivores and detritivores, each with many levels of specialization. When this trophic variety is associated with the many different microhabitats in water bodies that different beetle species may prefer, and the wide spectrum of body sizes they collectively span, assemblage composition can reflect a substantial number of environmental factors. However, Fairchild *et al.* (2000) noted that individual site successional influences had been studied rather little and, with the increasingly common conservation measure of constructing ponds to replenish lost habitat in wetland areas, this knowledge may be of value in planning management and thus be important to acquire. Contexts for such restorative measures include alleviation of wetland development, such as for highways or industrial or urban development, and measures such as reserve improvement. Comparative surveys of constructed and older (reference) ponds in Pennsylvania indicated that newly constructed ponds naturally acquired approximately the same number of water beetle species found in reference ponds, within a few years of colonization. However, most of these were relatively widespread species, and many of the rarer species occurred typically only at the older sites (Fairchild *et al.* 2000). Similar inferences have been made elsewhere. In Murcia, Spain, artificial systems (such as rivers influenced by dams, irrigation pools and channels) contained species of low conservation interest in comparison to those in more natural habitats (Sanchez-Fernandez *et al.* 2004).

Frequent colonizations may be necessary for many beetles to persist regionally, and early colonizing species can include large predatory Dytiscidae. Many aquatic beetles disperse readily by flight, with some species (*Dineutus assimilis*, Gyrinidae) recorded as flying more than 20 km (Nurnberger & Harrison 1995). An individual of the large, strongly flying *Hydrophilus piceus* (Hydrophilidae) has been reported from a North Sea oilrig, and Beebee (2007) suggested that it should be capable of colonizing newly created or restored sites quickly. This species persists in Britain in only six main breeding population centres.

One important clarification from Fairchild *et al.*'s study was to confirm the importance of fish impacts on beetle assemblage composition, with adult and larval water beetles both susceptible to fish predation. This observation was not new: the authors cited, for example, Weir's (1972) study in Africa, where a pool without the omnivorous catfish *Clarias gaveipennis* supported 31 species of beetles, compared with only seven in a pond with the fish, in which densities were also around 70 times less. Fish are probably universal predators of water beetles and, according to Fairchild *et al.* (2000), may overshadow the effects of other habitat variables on beetle assemblages. Six ponds with few or no fish supported nearly three times as much beetle biomass as their 12 ponds with substantial established fish populations, although only slightly fewer beetle species.

The major habitat division for water beetles is their association with running water (lotic) or still water (lentic), with rather few species inhabiting both of these equally. Thus, flow is sometimes assessed as the most important parameter in classifying water beetle habitats, but other factors are also very influential. For Dytiscidae, Larsen (1985) reported that other habitat variables influencing species distribution included acidity, elevation, presence/stability, salinity, size, mineral content, and several aspects of vegetation and detritus, with different habitats intergrading across these. Using this framework, a survey of 312 sites in Alberta, Canada (collectively yielding 145 species of Dytiscidae) was interpreted through cluster analyses based on species similarity to indicate 12 habitat-based

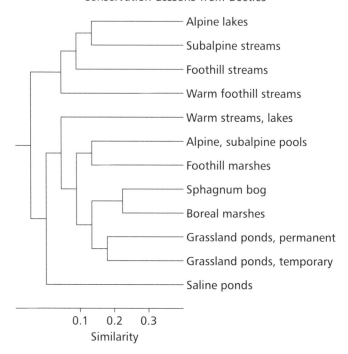

Fig. 9.1 Summarized graphical representation of the various habitats for dytiscid water beetles in Alberta, Canada based on similarity of species incidence. (From Larsen 1985 with permission.)

communities (Fig. 9.1). These varied substantially in richness, from 10 species in alpine lakes to 79 species in each of boreal marshes and permanent grassland ponds (Larsen 1985). Most species occurred most frequently in only one site cluster. Almost every species differed in its pattern of distribution across clusters, and very few were restricted wholly to sites within any one cluster. Despite the variety of detail, Larsen considered that his clusters demonstrated broad patterns of spatial and ecological distribution for major components of this diverse fauna.

Within the Irish fauna (165 species from 15 families), the most important variables differentiating the 10 assemblage categories distinguished by Foster *et al.* (2006) were flow, water permanence, exposure, and type of substrate, with acidity not isolated as a major determinant. Physical and structural components may generally be more important than chemical attributes, so that details such as bank profile may need to be considered in management. Considerable differences in water beetle assemblages may occur in steep-sided lakes and water bodies with shallow bank profiles (Nilsson *et al.* 1994), the differences possibly related to microclimate differences and provision of areas suitable for shallow-rooted plants to grow. As another possibility, a shallow margin may be easier to negotiate by larvae of species that leave water to pupate on land (Painter 1999). Habitat diversity was emphasized also by Lundkvist *et al.* (2001), who showed that diversity of Dytiscidae in Sweden depended on the variety of wetland types in a landscape, based on surveys in 12 wetlands in an agricultural landscape. Permanence of the habitat and the extent of shading were important influences, and a system for conservation of temporary, permanent and mixed-age or successional stages across both wooded and open environments was considered desirable for beetle richness conservation. Many Dytiscidae fly well, and can disperse over at least several kilometres to colonize water bodies. Comparative surveys using flight traps and water traps (Lundkvist *et al.* 2002) gave consistently higher species richness and abundance in the former. Urbanization may result in lower dytiscid richness than in agricultural landscapes. Site descriptions for water beetles vary considerably in complexity and, as for many terrestrial beetle sites, are sometimes difficult to harmonize for comparisons across studies. These vary very appreciably in the amount of water chemistry data, if any, included so that the detailed information include by Foster *et al.* (1990), for example, has not always been paralleled effectively.

In many places, particular beetles of conservation interest are associated with habitats such as brackish coastal pools, in which salinity may be an important environmental variable. Many coastal saltmarshes are flooded regularly, so that the salinity at any site is largely an outcome of the individual tidal regime and extent of isolation. As Greenwood and Wood (2003) pointed out, there has been rather little research on tolerances of aquatic Coleoptera to salinity changes over time, despite the large number of species found mainly in brackish water habitats. Saline lakes can support numerous beetles (Timms & Hammer 1988), with the most widespread species in Canada (*Enochrus diffusus*, Hydrophilidae) having perhaps the widest salinity tolerance range of any water beetle. Those of the related European *E. bicolor* are also broad. This species, one of several halobionts of conservation interest in Britain, was second only to *Rhantus frontalis* (Dytiscidae) in salinity ranges reported in Britain (Greenwood & Wood 2003). From the limited information available so far, increasing representation of Dytiscidae in brackish water assemblages seems to be related to decreasing salinity, but the optimal conditions for each of the many species involved remain unknown.

Local extinctions can sometimes be attributed directly to pollution, but widespread draining of wetlands in many parts of the world is a predominant threat to many taxa of water beetles and other aquatic invertebrates.

Ground beetles and tiger beetles

Perhaps more than any other group of terrestrial beetles, the ground beetles (Carabidae, the largest family of Adephaga) have played key roles in establishing appreciation of the importance of insects in conservation and as tools to establish habitat classifications for beetles. The tiger beetles, treated either as a subfamily (Cicindelinae) of Carabidae or as a distinct family (Cicindelidae), are a particularly notable flagship group, with conservation interests and impacts in many parts of the world. They are a spectacular and easily recognizable group, whose biology has been summarized by Pearson and Vogler (2001). With a few exceptions (e.g. *Gehringia* is only about 1.6 mm long; Maddison 1985), carabids are reasonably large and conspicuous beetles. Many species are active by day, and are therefore easy to observe and study. Carabidae are one of the larger families of beetles, with at least 25,000 species as estimated by Thiele (1977), later raised to 40,000 species by Lovei and Sunderland (1996), and the considerable variety within the family is reflected by their division into numerous (86) tribes (Lovei & Sunderland 1996). Local endemism is also generally high. The New Zealand fauna, for example, includes 50 endemic genera, of a total of 78 in the country (Larochelle & Lariviere 2007). Further, endemism may be reflected in proliferation of carabids in particular habitats, for example the abundance of those found beneath dehiscing *Eucalyptus* bark in Australia incorporates some 500 species, or approximately one-quarter of all those known from the continent (Baehr 1990). Many Amazonian species occur in the tree canopy (Lucky *et al*. 2002) (see p. 10). Carabids occur in all major terrestrial habitats and their assemblages are influenced greatly by patterns and changes in land use. In particular, surveys in Britain have demonstrated the importance of land cover to ground beetles (see Eyre *et al*. 2003a, and below). Their richness and widespread amenability to surveys by pitfall trapping has rendered them exceptionally attractive in environmental assessments, and considerable attention has been paid to their values as indicators (see p. 51). Suggestions of that value, however tentative, have been a major stimulus to their study. Large samples are relatively easy to obtain from assemblages, and many published studies enumerate several thousand individuals. They yield abundant information that assemblages change in relation to habitat characteristics and modifications. The value of local differences in assemblages of Carabidae has become apparent in many comparisons across habitats, and many of these differences have wider implications in landscape ecology. Thus, in Tasmania, ground beetle assemblages differ with uphill distance from streams in wet eucalypt forest, demonstrating the need for upslope areas to be included in reserves in addition to the more usual riparian areas that are, in many cases, more difficult to harvest and therefore pragmatically left alone (Baker *et al*. 2006). Carabids are also ecologically varied, so that ecological groups can be distinguished in assemblages, paralleling the functional groups noted earlier for staphylinids (see p. 48). However, feeding habits were only part of the character suite used to distinguish seven carabid groups, with Cole *et al*. (2002) employing 10 ecological attributes for this purpose (Table 9.1). Subsequent investigation of the pool of 68 species in Scottish farmland showed some groups to differ in representation in relation to intensity of management. Most of the groups transcended taxonomic groupings with constituents from more than one genus. However, the only group to parallel strictly taxonomic limits, that containing 'large *Carabus*' (five species), was also that most strongly influenced by agricultural practices, being absent or nearly so from sites where any intensive management occurred. This finding had been reported previously, possibly related to the long life cycles rendering them unable to respond to drastic or sudden changes associated with practices such as grass cutting, tillage and fertilizer applications, whereas smaller species with short life cycles are better adapted to cope with environmental fluctuations (Cole *et al*. 2002).

Many species are highly characteristic of particular biomes and thus have potential to indicate habitat quality in various ways. Very specific geomorphological features, such as large sand bars alongside rivers for some tiger beetles, may be

Table 9.1 Ecological traits of Carabidae, and the state attributes (total of 27), selected for use in multivariate analyses of assemblages to show combination of ecological, behavioural and morphological features of value in such analyses.

Ecological trait	Attribute state
Size (body length)	Very small (< 5 mm) Small (5–9 mm) Medium (9–15 mm) Large (> 15 mm)
Overwintering stage	Adult Adult and larvae, or only larvae
Duration of life cycle	1 year 2 years
Food of adult	Collembola specialist Generalist predators Mixed diet Mostly plant
Daily activity pattern	Diurnal Diurnal and nocturnal Nocturnal
Breeding season	Spring/summer Autumn/winter
Emergence of adults	Spring/summer Autumn
Main activity	Only spring/summer Autumn
Wing morphology	Apterous or brachypterous Dimorphic Macropterous
Locomotion	Runner Pusher Digger

Source: Cole *et al.* (2002) with permission.

correlated strongly with the presence of particular species that manifest such microhabitat specificity. However, lack of standardization in sampling renders most individual surveys difficult to compare. The substantial literature on ground beetle assemblages and their dynamics therefore largely represents a long series of independent surveys, with the responses of constituent species to the vagaries of the individual trapping characteristics usually poorly understood. As for many other insects, the spectrum of ground beetles captured differs very substantially across different techniques, so that survey conditions and objectives must be specified clearly and realistically, and sufficient details given in the final report to enable repetition by other workers. As one example, a broad comparison between pitfall traps and light traps in China over the same time intervals gave a pooled total of 33 carabid species in 18 genera (Liu *et al.* 2007). The number of species taken by each method was similar (pitfalls 24, light traps 23), but each method yielded a substantial proportion of captured species not taken by the other (pitfalls 10, light traps 9). With few exceptions the species taken in both sets of traps were strongly biased in numbers to one method or the other (*Chlaenius micans*: 167 in pitfalls, 4 in light traps; *Harpalus simplicidens*: 10 in pitfalls, 365 in light traps). Any attempt to conduct an inventory of carabids in an area must therefore heed such differences in 'trappability' and cater accordingly by using a variety of appropriate methods.

However, in parts of the northern hemisphere, carabid surveys have proliferated and become based on several relatively standardized approaches, so that habitat classifications owe much to incorporation of accumulated carabid data and reflect that many species are indeed highly characteristic of particular biotopes or associations. Thus, in examining their values for habitat classification in Britain, Eyre *et al.* (1996) used three datasets comprising (i) 194 species lists from exposed riverine sediments, mostly obtained by hand-collecting; (ii) 100 site samples from eastern Scotland from pitfall traps used in a standard manner; and (iii) 141 sites in Leicestershire sampled by standardized 30-minute hand searches. Analyses of these gave a 'sensible habitat classification' based on a wide array of variables, and Eyre *et al.* suggested that the substantial amount of information available on Carabidae over much of Europe could provide the basis for an integrated approach to cover a number of countries.

Many carabids are nocturnal, but others (including most cicindelines) are diurnal and conspicuously active. Most ground beetles, and all tiger beetles, are predators and at least part of the interest in their well-being results from some being manipulable predators of value in crop protection. Carabids are widespread, diverse and abundant in agricultural and forest ecosystems, and have received particular attention in the northern temperate regions. Many species are habitat specialists and very characteristic of particular biomes, so have gained a reputation as indicators of the quality of such environments and of wider influences such as climate change and land management practices, and as surrogates for appraising the impact of human activities. Thus, beach-frequenting cicindelines may be threatened by human recreational activities such as tourism and vehicle use, leading to local extirpations, with wider impacts on highly localized species. Larval and adult counts of one species, *Lophyridia concolor*, on beaches in Turkey indicated that crushing of larval burrows by trampling accompanied by possible movements of adult beetles to beach sections less heavily frequented by tourists might affect the beetle (Arndt *et al.* 2005). As noted earlier, tiger beetles may be very characteristic of their sandy substrate habitats, with larvae sometimes even more restricted than adults. The factors that influence the distributions of both adults and larvae of some coastal species, and the interactions between co-occurring species, are reasonably well understood. As one example, pairs of species in Japan in which adult mandible size is the same do not coexist, whereas species with different mandible lengths may have overlapping distributions (Satoh *et al.* 2003). The six species occurring in coastal

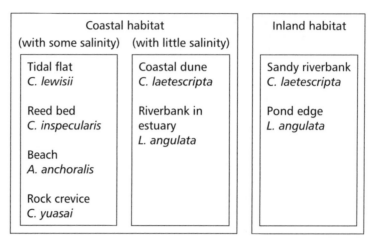

Fig. 9.2 Larval habitats of six species of coastal tiger beetles (*Cicindela lewisii, Lophyridia angulata, Chaetodera laetescripta, Abroscelis anchoralis, Callytron inspecularis, C. yuasai*) in Japan. (After Satoh *et al*. 2003 with permission.)

beach habitats collectively occupy a gradient in which several environmental factors change perpendicular to the shoreline, so that vegetation cover and diversity increase with distance from the shoreline and salt spray and sand grain sizes decrease, with the beach zone giving way to coastal dune. The micro-habitats occupied by larvae of each species (Fig. 9.2) may be related mainly to tolerance to salinity, with four species found at sites with some salinity and the other two only on non-saline sites (Satoh & Hori 2005), and inland. Tiger beetles have been promoted as a useful indicator group, and considerable additional biological and distributional information has accrued since Pearson and Cassola (1992) noted that the fauna of 129 countries was then relatively well known, a statement unusual for any large group of beetles. Writing on North American species, Pearson *et al*. (2006) noted that these collectively occupy 17 different habitat types, but that most species were each restricted to a single habitat category, and only a few species occurred in as many as six habitat types. Such high levels of definable habitat specificity are an important correlate of conservation interest, and some tiger beetles have become important flagships for specialized susceptible biomes that may lack other invertebrate ambassadors with comparable appeal. Also unusually, and important for conservation assessment, tiger beetle species richness in a given habitat/area can be estimated much more easily than that of many other insect groups. Pearson and Cassola (1992) estimated that only 50 hours on a single site may be sufficient to reveal 78–93% of species present, whereas the ensuing 200–5000 hours of sampling added only two to five species, these being mainly rare species near their range limits or arboreal species not necessarily amenable to direct searching techniques. Also importantly for surveys, students and others can be easily trained to recognize tiger beetle species far more rapidly than most other taxa, in which likelihood of taxonomic confusion is far greater.

Series of volumes on carabid biology (such as those edited by Erwin *et al.* 1979, den Boer *et al.* 1986, Stork 1990 and Brandmayr *et al.* 2000), resulting from periodic meetings of specialists, have provided very firm documentation of progress in much biological and distributional knowledge of ground beetles, while reviews (such as that by Lovei & Sunderland 1996) focus on more limited aspects. The proliferation of these beetles may largely reflect their major adult roles as actively hunting ground-dwelling polyphagous predators or omnivores (although some are herbivores), with a considerable array of ecological strategies and habitats available to them. Hammond (1990) noted their strength as being able to subsist on small and very scattered food items. Documentation of their biology in forest and agricultural ecosystems, predominantly in the northern temperate regions, has led to Carabidae becoming one of the best known of all invertebrate families and, in Niemela's (1996) words, 'exceptionally suitable for applied work; they are abundant, speciose and ecologically well-known', and they have become popular model organisms in conservation and land-use evaluation (Holland & Luff 2000). They are regarded widely as beneficial predators in cropping systems, and approaches to enhancing their use as agents in conservation biological control may transfer easily to wider conservation activities. The central themes of this endeavour focus on devising ways to increase the hospitality of agricultural systems (particularly of cropping areas) for polyphagous carabids valued as candidates in conservation biological control, so that both their abundance and their means of access to the pest species have practical importance in crop protection. Much related management involves increasing the 'associated biodiversity' (*sensu* Vandermeer & Perfecto 1985) in agricultural ecosystems and, in some instances, transforming it to managed biodiversity, with likely wider benefits for other taxa.

Much of this management within the agroecosystem arena has involved juxtaposition and interspersion of crops with contrasting more natural habitats, either natural remnants or near-crop areas created as permanently free of cultivation, notwithstanding occasional side effects (such as spray drift), but collectively providing refuges, more diverse and complex environments, that may also facilitate movements. The impacts of various crop husbandry practices on carabid populations have been studied extensively (Hance 2002). Predominantly, non-crop habitats incorporated in agricultural landscape management in Europe consist of three categories:

1   existing field margins, including grassy margins, hedgerows, constructs such as windbreaks, shelterbelts and fence lines, and riparian buffers;
2   margins established to prevent soil erosion or run-off of agricultural chemicals;
3   margins created to benefit game birds by providing shelter and a supply of arthropod food for their young – conservation headlands are perhaps the best example of these (Lee & Landis 2002).

In addition, within-field non-cropping areas have become significant considerations. The best known of these are beetle banks (Thomas *et al.* 1991, 1992), created as island habitats in fields with the express purpose of manipulating beneficial arthropods such as Carabidae. Beetle banks are simply raised strips, about 1.5–2 m wide, sown with grasses such as *Dactylus glomerata* (or other species, but avoiding any weedy species likely to contaminate the adjacent crops) and extending into crops. These structures have an establishment cost (reflecting their construction) and also take some land out of production, so clear benefits must be seen to accrue if they are to be adopted widely. Higher beetle densities than in adjacent areas have been found in beetle banks, and the advantages of having established predator populations well towards the centre of the crop may be considerable. One pioneering design of beetle banks specified a length of 290 m, and early evaluation reported these capable of harbouring as many as 1100 adult carabids per square metre over winter (Thomas *et al.* 2000). Surveys of botanical changes within beetle banks suggested that whilst plant diversity may increase, vegetation structure remains dense so that the banks continue to function as refuges for predatory invertebrates for at least a decade, with only minimal management (Thomas *et al.* 2002). The impact of predators on crop pests (such as cereal aphids on wheat; Collins *et al.* 2002) decreases with distance from the beetle bank.

The role of beetle banks emulates that of the long-established linear field margin, which also has the dual role of acting as refuges/reservoirs and affecting connectivity in the landscape. The latter includes considerations of 'permeability', in which studies of Carabidae have provided considerable understanding. Thus, studies on several carabid species in Norway showed that a grassy bank was less permeable than the adjoining barley crop (Frampton *et al.* 1995). In an earlier study (Duelli *et al.* 1990), a variety of carabids moved across the borders of a maize crop, and permeability to particular species depended on features of the margin. Mark–release–recapture surveys of three abundant species (*Harpalus rufipes, Pterostichus melanarius, P. niger*) showed that (i) grassy banks (either 60 cm or 1.2 m wide) slowed recaptures in relation to a barley crop; (ii) hunger of the beetles (assessed by comparing adults that were starved or well fed before release) affected dispersal rate, as did the bank; and (iii) interspecies differences of detail occurred. Although the causes of lower permeability of the grassy bank were not investigated directly, Frampton *et al.* listed several possible contributing factors, namely (i) changes in movements with microclimates and density of vegetation; (ii) relatively high prey density in the grassy bank areas; (iii) differences in beetle burrowing behaviour between the crop and bank; (iv) the topography of the bank, with steep slopes being avoided by beetles; and (v) presence of a bare soil strip between the crop and the bank. Any or all of these factors may need to be investigated in clarifying such effects, but the overall results from this study suggested consistent differences in cross-margin movement as a consequence of landscape manipulation. One very pertinent comment relating to the permeability of field boundaries merits reiteration in all related contexts: 'no matter how impermeable the boundary might be, all fields have gates or bridges for access. If tractors can pass between fields, it is reasonable to assume that beetles are also able to do so' (Thomas *et al.* 2002, p. 333). Gaps in hedgerows for vehicle passage are typically 7–9 m wide, and are readily crossed by beetles such as *Nebria brevicollis* (Joyce *et al.* 1999). This carabid inhabits both boundary and crop environments, and is an important predator of aphids and has relatively high dispersal ability. As emphasized earlier (see p. 75), generalization is difficult because features such as roads may substantially restrict movements by some species.

The structure of field margins can strongly influence carabid richness and well-being. As another example, Asteraki *et al.* (1995) showed that grassland field margins with a hedge supported considerably more species (63) than those with a post-and-wire fence alone (48). The conservation roles of grassy strip margins had been little studied before the 1990s (Kromp & Steinberger 1992), with most attention by that time addressing their function as refuges for predators. For this, and for wider conservation issues, the complexity of field margins is of critical importance, so that the more structurally diverse environments of hedges, for example, perhaps provide greater overwintering success than simple grass edges (Varchola & Dunn 2001) and also support greater supplies of prey.

The differences between a corridor and a barrier to beetles are sometimes very subtle, and these influences include the characteristics of the edge. Studies on the carabid *Abax parallelopipedus* by radiotracking individuals (Charrier *et al.* 1997) revealed effects on beetle behaviour related to the abruptness of the transition between the edge and the crop, implying that land-use pattern may be an important determinant of beetle opportunity to disperse. A more gradual transition (softer edge) was reasonably permeable, but a more contrasted (harder) edge posed a barrier to movement. As Bommarco and Ekbom (2000) noted for *Pterostichus cupreus*, an individual beetle is likely to encounter a variety of different habitats during its lifetime, and the nature of the crop or intercrop boundary may itself change considerably over that period. For example, as crops grow, the microclimate and availability of food may also change, so that a beetle's responses to an edge may change over a growing season. The habitat edge (barley crop/ley) clearly influenced movements of *P. cupreus*, as shown also by diffusion models (Bommarco & Fagan 2002). This edge was assessed as soft, and thereby of positive value for movements of the predator into the crop. This study revealed the practical importance of understanding the beetle's movement behaviour, and the limits posed by some experimental approaches. *Pterostichus cupreus* normally moves asynchronously because of each beetle having distinct resting periods, so that natural population-level movements may be considerably less than implied from mass-release studies through which such individuality is removed. A further practical inference is the need to foster early-season movement from edges to crops to suppress pests, implying that agricultural landscape management may need to appreciate such species movement patterns and seasonal foraging behaviour as factors in manipulations of land use.

Carabidae were among the taxa used to formulate a general framework of how insects can move through and between habitats within agricultural landscapes. Other than for large-scale migrations bringing species into the area, five major classes of movement have been distinguished (Wratten & Thomas 1990).

1. Colonization of crops from non-crop habitats (spring) and reverse movement from crops to non-crop areas (autumn).
2. Movements between different crop types or stages during the growing season, in response to different microclimates or food needs/availability.

3 Colonization of new habitats as the area under intensive agriculture expands and contracts.
4 Recolonization of land on which pesticides have earlier reduced numbers, as their effects decline over time.
5 Numerical responses to areas particularly suitable within the crop, such as areas of high prey density.

These collectively emphasize the importance of understanding landscape influences on any focal species selected for management. The categories also indicate the variety of temporal scales needed for changes, from short term to across seasons or years.

The multitude of studies investigating changes in carabid assemblages with environmental variation and across vegetation boundaries has led to a far better understanding of dispersal capability and potential than is available for most other families of beetles. Dispersal ability and opportunity varies widely among carabids, and many species are flightless or exhibit some form of flight polymorphism. Many carabids apparently move rather short distances (Thiele 1977; Luff 1987), whilst others disperse extensively. The incidence of such varied species or conditions in assemblages may indicate some aspects of likelihood of colonization, and provide important guidance to species conservation and to sampling. If it is initially presumed uncritically (and incorrectly) that all species present can fly, some trapping systems may prove highly unsuitable for inventory studies. In general, extended flight, even in macropterous ground beetles, appears to be rather unusual (den Boer 1971). Meijer (1974) distinguished five categories of adult beetle in his study of colonization of new polder land by carabids.

1 Macropterous species that can fly at any stage of adult life.
2 Macropterous species that fly mainly whilst young, but are still capable of flight when they become older.
3 Macropterous beetles that fly only when young, after which their flight muscles autolyse in parallel with reproductive development.
4 Dimorphic or polymorphic species, in which macropterous individuals can fly. In this group, flight muscles may autolyse in some species but not in others, so that macropters fit also into one or other of the previously mentioned groups.
5 Brachypterous species which are almost always flightless, and presumed not to fly.

In Meijer's (1974) survey, only about half the carabid species were caught in window traps, rather than on the ground, and are thus the minimum suite of species that are proven able to disperse by flight, and dissection of females revealed that many of the early arrivals were reproductively immature.

Dung beetles

The comprehensive summary of dung beetle biology edited by Hanski and Cambefort (1991) includes much foundation information to emphasize their value in conservation. In his introductory chapter, Hanski (1991a) emphasized the patchy and ephemeral nature of their key resource, dung, and that up to 'thousands of individuals and dozens of species' may be attracted to single droppings in parts of the world. Dung beetles are colonizers and exploiters par excellence, and many are ecological specialists whose presence is determined by features of the dung habitat, and whose incidence and abundance may vary seasonally. The ecological discussion and interpretation in *Dung Beetle Ecology* underpins many considerations of conservation biology, yet the term 'conservation' does not appear in its index. Nevertheless, the recent uses of scarabs as ecological indicators and in various forms of ecological assessment stem firmly from the premises discussed there.

Dung beetles have become popular and important experimental vehicles for investigations of ecological processes and the interactions between species and of the factors that influence their richness and abundance. A collective group of scarab specialists has formed the Scarabaeinae Research Network (ScarabNet) dedicated to developing the use of these beetles as a focal taxon for studies in invertebrate biodiversity. Part of the attraction of dung beetles for ecological studies has been the advantage that enormous numbers of individuals and species can be collected easily by simple methods, and that their ecological impacts can be demonstrated easily. Consider two examples cited by Hanski (2005).

1   For cow pats in Finland, samples of 62,500 beetles included 179 species, extracted by dropping the dung pats into a bucket of water and submerging them by pressing down with chickenwire mesh, following which the beetles leave the dung and float to the surface (Hanski & Koskela 1977).

2   Anderson and Coe (1974) found that 16,000 individual beetles arrived at a 1.5-kg heap of elephant dung in Kenya, and ate, buried and dispersed this completely within 2 hours. However, they also noted that beetle activity was much lower during the dry season, when dung piles could persist for 2 weeks.

The latter study reflects that African savannas support a greater variety of grazing/browsing mammals and higher dung beetle diversity than anywhere in the world, so that substantial ecological studies can be made. In another example, Tshikae *et al.* (2008) obtained more than 69,000 individuals, representing 67 species, in only two 24-hour pitfall samples, using a total of 25 traps each day. The most significant factor promoting their value in ecological assessments has thus been the ease with which assemblages can be studied and sampled by using baited traps, and by which captures can be compared across sites and habitats, with little variation in sampling materials or effort, to establish their patterns of incidence. These patterns have included range expansions, some due to deliberate introductions of species and others where a number of native dung beetles have benefited from human activities in that the availability of large herbivore dung (from cattle) has increased in many places (Hanski 1991b). Several native Australian dung beetles using bovine dung are now probably more abundant than at any time before European settlement. In addition, some genera and major functional categories are widespread, so that comparisons, for example, of the faunas of *Aphodius* in Europe and North America are realistic characterizations of regional differences (Table 9.2). Dung beetle taxonomy is

Table 9.2 Ecological comparison of the faunas of *Aphodius* dung beetles in Northern and Central Europe and eastern North America.

	Northern and Central Europe	Eastern North America
Approximate total number	75	50
Pasture species	48	8 + 12*
Forest species	5	20
Specialists†	15	10

* Introduced pasture species, most from Europe.
† Specialists in Europe include 10 montane species, many of them pasture species. Others are saprophages or coprophages with special macrohabitat/resource needs.
Source: Hanski (1991b) with permission.

generally well advanced, so that a high proportion of species captured can be recognized easily, and named. Three main functional groups were distinguished by Hanski and Cambefort (1991).

1   Rollers: the classical image of the sacred scarab with a ball of dung, exemplified by many Scarabaeidae.
2   Tunnellers: burrow beneath the dung and transport it to the burrows, as in Geotrupidae and some tribes of Scarabaeidae.
3   Dwellers: simply eat through the dung without forming any burrows or balls, exemplified by many Aphodiinae, including the large genus *Aphodius* with some 1650 species.

Within assemblages, some genera have become prolific, and afford an opportunity for studying the niche relationships and ecological, temporal and distributional overlap of closely related taxa. As one example, again from *Aphodius*, the 36 species found in Finland fall broadly into both seasonal and habitat-based guilds. Thus, the early summer cohort of species includes four such guilds in relation to the major biome in which they occur: forest specialists (reproducing mainly in forest habitats), generalists (breeding and feeding about equally in shaded and open habitats), intermediate species (clearly more common in open areas, but found in both categories), and pasture species (found more or less wholly in open pasture areas) (Roslin 2001).

Doube's earlier (1990) functional classification of dung beetles was extended by Finn and Gittings (2003), who emphasized the likely value of this approach in seeking ecological generalities, particularly in the wider roles of competition, and extending from more limited individual studies. They demonstrated the considerable variety of important influences on dung beetle richness (using northern temperate communities), transcending many resources and affecting the immigration of beetles to dung pads (as summarized in Fig. 9.3), across a range of temporal and spatial scales (Fig. 9.4). Each of these many factors may need to be considered in a species or assemblage management plan, and the principles from this regional survey extend easily to other places. One important study aspect suggested by Finn and Gittings, in noting that most research has

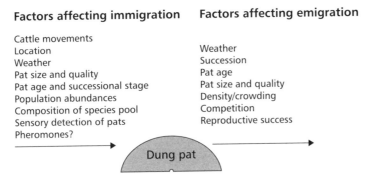

Fig. 9.3 Major processes and factors that affect immigration and emigration of dung beetles to and from dung pats in pastures. (From Finn & Gittings 2003 with permission.)

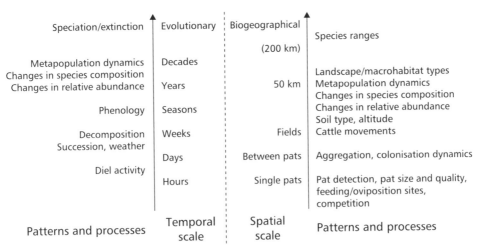

Fig. 9.4 Integration of community ecology of dung beetle assemblages indicating the processes and factors that can operate at different temporal (left) and spatial (right) scales. (From Finn & Gittings 2003 with permission.)

focused on adult dung beetles, is the need for further studies on interactions between larval stages, among which competition both within and between species is likely to be more potent than in adults.

As a consequence of numerous surveys, the distributional patterns of many dung beetles and their functional groups are not only reasonably well defined but the factors limiting those distributions are also understood. In addition to dung type (reflecting distribution of donor mammals), soil type may markedly influence burrowing species, and climatic regimes impose limits on many taxa. Although most species are restricted to mammal dung, some New Zealand beetles feed on bird dung, and the Australian *Cephalodesmius* forms balls of

Beetles in Conservation

Table 9.3 Numbers of species and individuals (proportion of total in parentheses) of dung beetles trapped at five kinds of dung/carrion bait at Chobe National Park (Botswana) in two 24-hour periods.

Bait	Species	Individuals
Cattle dung	52 (78%)	16,393 (23.7%)
Pig dung	51 (76%)	38,670 (56%)
Elephant dung	44 (66%)	7,773 (11.3%)
Sheep dung	37 (55%)	5,564 (8%)
Carrion	22 (33%)	675 (1%)

Totals comprise 69,075 individual beetles in 67 species.
Source: data from Tshikae *et al.* (2008).

chewed vegetation, perhaps reflecting the origin of the dung-feeding habitat thorough saprophagy. Within the spectrum of mammal dung as a key resource, considerable specialization can occur, sometimes to single or a few related host species. For example, some species of *Uroxys*, in South America, are restricted to sloth dung. In Australia, several species of *Onthophagus* found with wallabies occur on the host, congregating around the anus and grasping the fur. As faecal pellets are expelled, the beetles drop to the ground with them, and commence to process the fresh dung. Such preferences have been studied on a number of occasions. Many of the 30 most abundant dung beetles in woodland savannah in Chobe National Park, Botswana specialized on different dung/carrion, although the most abundant species were relatively generalized and attracted to all five baits tested (Tshikae *et al.* 2008). The five baits used and the numbers of dung beetles obtained in only two 24-hour trapping periods, each involving five pitfall traps with each bait, demonstrate clear differences within the overall pool of 67 species (Table 9.3).

Short-duration sampling of assemblages, as emphasized in several of the above-mentioned studies, and numerous others cannot reveal the patterns of seasonal incidence and distribution of all species in an area, so that interpretations of assemblage composition and dynamics must be correspondingly cautious. Two main categories of factors influence such interpretations: (i) limited seasonal periods of apparency or incidence, so that the beetles are present for only part of the year, and (ii) differing seasonal availability of resources, here dung. The distinction between these effects may not initially be clear for any given case. To exemplify the variation that can occur, assemblages of scarabaeoid beetles in Spain could be classified into eight groups on combinations of habitat and season (Galante *et al.* 1991), with abundance of some representative species indicated in Table 9.4. Some species are present for much of the year, whilst others are more seasonal, and their relative abundance in woodland and open (pasture) habitat ranges from consistent (*Onthophagus ruficapillis* was more abundant in open habitats) to highly oscillatory (*O. similis* changed in habitat predominance over the same period). Different species made the greatest

Table 9.4 Seasonal and habitat separation of dung beetles near Salamanca, Spain: numbers of individuals trapped each season in wooded habitats (W) and cleared pasture (O).

	Spring		Summer		Autumn		Winter	
	W	*O*	*W*	*O*	*W*	*O*	*W*	*O*
Gymnopleurus flagellatus	–	–	–	6	–	–	–	–
Copris lunaris	72	21	418	13	12	1	–	–
Copris hispanus	11	15	32	–	15	4	–	–
Chironitis hungaricus	–	–	–	4	–	–	–	–
Onitis belial	–	–	1	5	–	–	–	–
Bubas bubalus	101	141	402	162	94	35	1	1
Euoniticellus fulvus	–	8	39	71	2	6	–	–
Caccobius schreberi	–	26	38	192	–	–	–	–
Onthophagus taurus	–	11	6	24	–	1	–	–
Onthophagus punctatus	–	1	1	–	1	–	–	–
Onthophagus furcatus	–	4	25	3	–	–	–	–
Onthophagus ruficapillus	3	480	30	172	–	–	–	–
Onthophagus coenobita	5	–	–	–	–	–	–	–
Onthophagus similis	477	1050	1003	294	289	562	5	86
Onthophagus vacca	9	52	51	73	–	5	–	–
Typhaeus typhoeus	12	2	–	–	24	–	11	–
Geotrupes mutator	22	2	11	5	54	7	14	1
Geotrupes ibericus	2	1	29	1	35	3	–	1

Source: Galante *et al.* (1991) with permission.

contribution to differentiation between woodland and pasture assemblages and between summer/autumn assemblages in pasture. Some changes (for example that of *Bubas bubalus*) may reflect movement in response to microclimate, in this species from pasture to more sheltered woodland in late spring. Paralleling some other observations, most of the predominant woodland scarabs are larger beetles, and most of those in open habitats are medium-sized or small, with the difference possibly reflecting microclimate influences.

Australian tropical dung beetle assemblages can change over very short distances (Hill 1996) and seasonality can be pronounced, with most species found only in the wetter months. However, seasonality differed between *Eucalyptus* woodland (with beetles mainly in the late wet season and early dry season) and wetter adjacent *Allocasuarina* forest (with most beetles in the late dry season and early wet season) (Vernes *et al.* 2005). These authors used bait of fresh dung below elevated cage-held native mammals. With catches corrected for differing numbers of mammals captured across seasons, beetle numbers were significantly higher than expected in the late wet season, and significantly lower than expected in all other seasons, but these numbers were biased heavily by one common species found in both vegetation types, whereas the other 10 species were all associated strongly with either *Eucalyptus* (six species) or *Allocasuarina* (four). The exception, *Coptodactylus subaenea*, comprised 409 of the total 489 beetles (301 in *Eucalyptus*, 108 in *Allocasuarina*).

Within the diverse South African fauna, Koch *et al*. (2000) assembled records of 509 species in 59 genera to examine the presence of dung beetles in protected areas. Their analysis suggested that the habitat of 166 of these was protected in existing reserves, with a more generous analysis (including mapping grids, each 25 × 25 km, with as little as 25% of their area protected) increasing this to 366 species (72% of the total). Although the South African reserves network has been planned largely by needs for mammal conservation, dependence of many dung beetles on herbivore dung has enabled many of these species to co-survive. Five possible mechanisms were advanced to account for this relatively high dung beetle representation in conservation areas.

1   Human activity artefact 1: the continued survival of non-ruminant herbivores (such as elephant, rhinoceros, zebra) in reserves, with their large-scale disappearances from the wider landscape.
2   Human activity artefact 2: continued presence of a variety of game species (especially browsers) in reserves compared with their absence and large-scale replacement by grazers (mainly cattle) outside reserves may have helped to facilitate conservation of some specialist feeders.
3   Generalist dung feeders are likely to occur in a broad range of habitats and are thus likely to be sampled by reserves wherever these occur.
4   Collection artefact 1: much of the accumulation of intensive reference collections has occurred in reserves, where a high diversity of dung beetles was anticipated to be present. Known high-diversity areas have sometimes been targeted for such surveys, and may have introduced bias into the overall data available.
5   Collection artefact 2: linked with this, other areas have been undersampled, so that (particularly for the more arid regions) the fauna may have been inadequately sampled.

All of these may contribute to the pattern, but the unknown effects of the two collection artefacts are those that most hamper sound conservation decisions. Koch *et al*. (2000) suggested that their effects may be large, and render selection of additional optimal areas for conservation based on dung beetles difficult.

Many dung beetles have been introduced deliberately to areas where they do not occur naturally, as alien species that then interact with native species directly as invaders. Such situations have considerable relevance in species conservation, and the continuing importation of dung beetles to Australia for control of cattle dung provided the opportunity to address three relevant themes (Doube *et al*. 1991).

1   Can successful invaders be predicted and selected?
2   Can predictions be made as to which communities will be invaded most easily?
3   Can the impacts of the invasive species on the communities they enter be predicted?

The factors seen to be significant in establishment are climate matching, habitat matching and reproductive biology, in conjunction with the numbers and modes of initial release. The variety of climate influences, as estimated through the CLIMEX model, were displayed effectively by Tyndale-Biscoe (1990) in her maps indicating the likely variations in performance of introduced dung beetles in different parts of Australia and so guiding introductions towards success. Likewise, species associated with particular vegetation or soil types in their areas of origin may fail in other environments when introduced. For example, in Australia, extensive use of superphosphate fertilizer on dairy properties may reduce the likelihood of some species establishing. In general, successful introductions were of multivoltine high-fecundity species, for which the outcome becomes clear within a few years. Many of the more specialized low-fecundity species may take much longer to investigate, with records of some extending over a decade or more before the outcome is clear. Because of the scarcity of native dung beetles in pastures in Australia, introduced species encounter little competition from these. However, they may encounter other introduced species, so that competition may then occur. Outcomes are difficult to anticipate, and may differ widely in different parts of the range of each, to incorporate climate influences.

The major aim of the introduction exercise in Australia, namely to increase the rate and extent of dung dispersal and removal of breeding sites for flies, appears to have been widely successful. Numbers of predatory insects in dung pats also decline with beetle presence, and the numbers of introduced dung beetles have led to them dominating cattle dung communities over much of Australia, so that those communities now centre on alien species. Their impacts are probably relatively low in more enclosed forest and woodland environments.

The presence and continued supply of dung from large herbivores is critical for many dung beetles, whether this is natural or supplied by surrogates such as domestic stock. Proliferation of cattle dung, although widespread, is only one example of change of a natural resource for scarabs resulting from human activity. For example, urbanization leads to an increase in dog dung, which (in common with that of other carnivores) is not exploited widely by native dung beetles, probably because of the low cellulose content (although this is somewhat overcome by the high fibre content of commercial pet foods). Dung beetles are gradually becoming more widespread in dog dung in Australia (Faithfull 1992), India (Oppenheimer & Begum 1978) and Europe (Carpaneto *et al.* 2005). A study in Rome focused on dogs replacing earlier livestock grazing to become the only large mammals in the area studied. Comparison of the dung beetle faunas in 1986 (sheep droppings) and 1999 (dog dung) indicated both lowering of species richness (19 to 9 species) and increased abundance (Carpaneto *et al.* 2005). Comparison with other areas, not dominated by dogs, in 1999 suggested that this change in food resource had fostered change in species composition to a greater extent than had either patch size or habitat quality. In addition, dog dung might provide a temporary refuge for some species that might otherwise become locally extinct in urban environments. In particular, a few species adapted to the dung of omnivorous mammals may be supported by dog dung. As well as the supply of dung being patchy, the amounts and dispersion

reflect the abundance and habits of the mammal providers, so that some deposition patterns may be characteristic. In South America, Horgan (2005) noted
that defecation patterns vary considerably among species, reflecting their size,
digestive physiology and social and foraging behaviour so that, for example, howler
monkeys defecate more or less simultaneously once in the morning and again
after a resting period during the middle of the day, to produce large aggregations of dung piles. Regular use of roosting sites may also produce local dung
concentrations.

Many workers have emphasized that local distribution of dung beetles may
be influenced not only by the dung resource but also by vegetation cover and
soil type. There may thus be defined and consistent differences between assemblages in forests and more open country (Hill 1996). Widespread parallels
have suggested the value of dung beetles as indicators of local environmental
changes. Clear associations of species with different forestry regimes in Sabah,
Malaysia (Davis *et al.* 2001) indicated that some species responded to disturbance. In Mexico, clearcutting of forests affects most species living within
the forest, but those species living near forest edges can adapt most easily to
that form of disturbance (Halffter *et al.* 1992). Scheffler (2005) distinguished
between disturbance-sensitive species and moderately disturbance-sensitive species
as well as disturbance-adapted species, but only 10 of her 60 Neotropical species
varied significantly in abundance and/or biomass between forested, selectively
logged and clearfelled areas. Six occurred predominantly in intact or selectively logged forest (low disturbance), and none of these was found in pasture
or clearcut areas.

In a wide overview of effects of forest fragmentation on dung beetles, Nichols
et al. (2007) surveyed the conditions of beetle richness and diversity with a
variety of environmental variables as expressed across 33 published studies in
an attempt to discover generalizations in community responses.

1   The richness and abundance of dung beetles in intact forests is very similar
    to that in modified forests, such as selectively logged forest and secondary
    forest, with high tree cover.
2   Heavily modified areas with little tree cover have only low-richness com-
    munities, sometimes with a preponderance of individuals of small-bodied
    species.
3   The decline in intact forest species richness and abundance is often accom-
    panied by increases in species characteristic of more open habitats, with the
    extent of this reflecting the wider local landscape context of the fragmentation.

The extent of isolation of the fragments may also be influential, but many
other factors also likely to be important (such as dung availability, changes in
vegetation structure, time since fragmentation) need further investigation.
Importantly, Nichols *et al.* cautioned against extrapolating to generalities about
the influence of dung availability, because the entire mammal community is
usually inadequately documented, together with its own responses to the
modifications being considered for beetles. Thus, comparison of intact forest
beetles with those in tree-dominated plantations in Borneo (Davis *et al.* 2000)

showed species diversity to be much lower in all plantation sites than in primary rain forest, with the plantation species a subset of the primary forest taxa, mainly of edge-habitat species. This study also evaluated the relative incidence of Borneo endemic species and more widespread beetles to establish whether the endemics might tolerate disturbance. The plantations contained both endemic and widespread species, although no forest-interior endemic species were found there. The spectrum of mammals present as dung donors can be influenced strongly by deforestation, which also affects the dryness of the ground, with this combination of food and soil texture critical to many dung beetles. However, the relationships can become complex. Debarking of conifers by Sika deer (*Cervus sika*) in subalpine Japan, with consequent death of the trees, led to increased grassland and loss of diversity in the resident dung beetle assemblage (Kanda *et al.* 2005). Although the increasing deer population increases the supply of dung, this has not increased the abundance or diversity of the beetles, which appear to be influenced more strongly by ground moisture, in turn related to exposure.

Agricultural systems, with the substantial ecological changes they engender, are sometimes implicated as threats to dung beetles dependent on dung of native mammals or in forest habitats and unable to thrive in the alternative environment presented by domestic stock on pasture. However, some more recent trends associated with abandonment of farming in parts of Europe can have converse effects. Large parts of some regions, such as the Maritime Alps and Appenines, previously grazed by large herds of cattle and sheep are now only sparsely grazed and are undergoing natural afforestation, with substantial such reversion since the 1980s. In north-west Italy, most of the 27 species of Scarabaeoidea trapped by Barbero *et al.* (1999) were significantly associated with only one of the four categories of dung offered as bait (namely cattle, deer, horse, wild boar), and with either woodland or open habitats. The general conservation implication again emphasizes the importance of retaining habitat heterogeneity to provide support for a variety of herbivorous mammals, and a variety of macrohabitats. In agricultural environments, remnant patches of woodland may be needed to support the naturally occurring array of scarabs, and in woodland environments patches of open ground may be needed for the same purpose, with the principle important also for reforested areas intermingling with cleared pastoral areas. The conservation inferences are paralleled in South Africa, where one important lesson from comparison of human- and elephant-induced disturbance (see p. 53) is that concerns arise from the management of large herbivores inside protected areas, in addition to the more usually acknowledged effects of habitat conversion outside such areas (Botes *et al.* 2006).

Related species, even when sharing the same resource, may show substantial ecological differences and manifest different population structures, so that members of an assemblage may respond in different ways to any given change. One consequence is that interspecific differences contribute to community composition in part as a function of landscape effects on the connectivity and well-being of each. *Aphodius* dung beetles feeding on livestock dung in pastures may thus constitute different assemblages depending on the distribution of pastures in the landscape (Roslin & Kiovunen 2001). Some species appeared to manifest different patterns of persistence over a landscape: *A. fossor* showed a largely patchy population, whereas *A. pusillus* exhibited a classic metapopulation structure, with movements between local populations much more frequent in the former species, and the latter being the least mobile of the 10 *Aphodius* species examined. *Aphodius fossor* is still widespread and abundant in Finland, but distribution of *A. pusillus* has recently declined in parallel with habitat decline.

Stag beetles

Stag beetles, Lucanidae, include some of the most spectacular and imposing of all beetles, many of them characterized by having enormous mandibles, and are highly desired by collectors. Durer's engraving of a male European stag beetle (*Lucanus cervus*) is one of the most enduring images of any beetle, but countless depictions of stag beetles reveal many such remarkable images. Some other lucanids are not as spectacular, but their appeal to collectors then flows from their rarity and local endemism. Thus the New Zealand fauna of 39 species includes 35 endemic species in five endemic genera (Holloway 2007). Most of these are relatively inconspicuous, not highly coloured and many are rather small and flightless. Holloway noted that two flightless species of *Geodorcus* have been given legal protection in New Zealand to help prevent over-collecting, but considered this measure desirable for all flightless species of endemic insects there. Few of these species are widespread, and several (six) are known only from single mainland localities, from only either North Island or South Island, and some (four) only from offshore islands.

Lucanidae are characteristically forest insects, associated with saprophagous wood-feeding as larvae, and are widespread. Larval development can take several years. With around 1250 species, the family is not particularly large, but many species are regarded as threatened by over-collecting and forest felling and by associated processes such as removal of dead wood and habitat fragmentation; some are important indicators of forest quality as members of the diverse communities of xylophagous or saproxylic insects that have long been of conservation significance (Speight 1989). Particular taxa of Lucanidae have become threatened and thus conservation targets in many parts of the world, and considerable efforts have been made to assemble information on the conservation needs of this family, beyond impacts of collector activity (see p. 118).

Tasmania has around 30 species of Lucanidae, and several endemic stag beetles there have recently received considerable conservation attention because some localized species occur in forests scheduled for clearing and conversion to plantations as a consequence of increased forestry activity. Similar operations, of course, occur in many parts of the world, but many others cannot be refined or regulated for conservation of insects to the extent that might prove feasible in Tasmania. Similar losses of beetles are projected to occur whether the plantations replacing native forest are of monoculture eucalypts or pines (see p. 151). Members of two lucanid genera, *Lissotes* and *Hoplogonus*, pose particular concerns. The flightless broad-toothed stag beetle *Lissotes latidens* occurs in moist eucalypt forest, with the extent of suitable habitat estimated as 43 km² (of a total range of 280 km²) in south-eastern Tasmania, with a portion of this habitat on Maria Island (see Clarke & Spier 2003 for summary). Much of the mainland range is restricted to riparian areas, and so is very patchy and fragmented. Larvae and adults both occur in upper soil layers beneath rotting logs, and *L. latidens* is apparently soil-dwelling throughout its life, as are a number of related species. Despite lack of quantitative population data, densities are thought to be lower than those of many other Tasmanian forest lucanids. Both *L. menalcas* and *Hoplogonus simsoni* occur in similar areas, but perhaps are not as much at risk as *L. latidens*, which is listed as endangered and is legally fully protected. The listing conditions oblige developers to investigate its presence in sites proposed for forestry operations, but the major need is protection of habitat in conjunction with sympathetic forest management. Maria Island is largely a national park, and contains around 12% of the beetle's known habitat (Clarke & Spier 2003), but most mainland populations are not in comparably secure reserves, although some small riparian reserves are exempt from logging or felling.

The presence of *L. latidens* is strongly correlated with coarse woody debris (Meggs & Munks 2003), but determining the major correlates of resource/ habitat needs in such rare species is often problematic. In their extensive direct searches for this beetle (by rolling logs and searching beneath them, and breaking open a subsample), Meggs and Munks found only 54 adults: 46 under fallen dead wood, five wandering on top of leaf litter or roads, and three identified from old body fragments in litter. Nevertheless, it occurred at sites with a variety of different disturbance regimes, including sites burned within the previous 50 years. Generally, only one or two specimens were found at any site, with the highest number at any site (six) translating to an adult density of only 30/ha. The possibility that very large fallen logs could provide refuges for greater numbers of beetles was at least partially refuted by Grove *et al.* (2006), who used an excavator to roll over or shift (and, after examination, replace) a combined length of nearly 900 m of logs to reveal only two living individuals of *L. latidens*. The low densities throughout surveys suggested that hotspots of abundance as potential conservation foci do not exist, so that selection of priority areas for reservation or protection could not justifiably invoke beetle numbers. Conservation then focuses on existing reserves and planning to ensure that populations on non-reserved areas do not become isolated.

The other species from Tasmania that has been studied in considerable detail is Simsons stag beetle *Hoplogonus simsoni* (see p. 151), which is found over an area of about 250 km² in north-eastern Tasmania, within which suitable habitat comprises 174 km². *Hoplogonus simsoni* needs a cool moist environment and areas from which fires have been absent for some time. It is distributed very patchily, with high-density populations only towards the eastern parts of its range (Meggs *et al.* 2003), and few beetles occur in mixed forest or rain forest; dry eucalypt forest is unsuitable and untenanted. It was not found in areas converted to pine plantations, but did occur in wet eucalypt forest allowed to regenerate after clearfelling. The optimal habitat, characterized through development of predictive modelling (Meggs *et al.* 2004), is poorly reserved and much of it targeted for conversion to pine plantations. *Hoplogonus simsoni* appears to be a late-succession or mature forest specialist. Again, conservation measures focus on limiting impacts on the beetle of plantation establishment, with a three-tiered strategy involving (i) proposing some areas of optimal habitat as wildlife priority areas, (ii) maintaining links between these and non-reserved important areas, and (iii) managing forestry activities by prescription in habitats throughout the rest of the range. Although some of the beetle's range is included in reserves, much of that is regarded as only marginal habitat (Meggs *et al.* 2003). The second parameter reflects the likely need for movement of beetles from source areas to enhance chances of long-term survival. However, individual beetles are believed to be able to move only 100–200 m over their lifetime. The third tier endorses the second, in regulating disturbances to assure sufficient regeneration time for *H. simsoni* habitat on a landscape mosaic basis. The detailed distributional mapping undertaken by Meggs and his colleagues is important also in detecting the places where interaction between conservation need and forestry activity are potentially likely to be most intense. The three predictive models developed by Meggs *et al.* (2004) to help determine the areas varied in performance, but the entire exercise is one of very few suites of predictive distribution and abundance models for a threatened insect that have been subsequently field-checked and evaluated.

The principle that underpins conservation interest in these Tasmanian lucanids stems from legal requirements for listed species to be taken into account during planning and implementing any forestry operations, so that threatened fauna, including invertebrates, must be catered for by appropriate management and objectives in areas subject to production forestry, by measures based on sound understanding of the species and specific threats to them. Two further species of *Hoplogonus* are also significant: conservation measures, based on survey and habitat characterization in north-east Tasmania, have been discussed for *H. bornemisszai* and *H. vanderschoori* (Munks *et al.* 2004). They are listed as endangered and vulnerable respectively, and at the time these species were listed (1999) lack of knowledge hampered development of sound management prescriptions. Although the ranges of these two species do not overlap, that of *H. bornemisszai* is largely contiguous with that of *H. simsoni*, and only small parts of the range of each are in reserves or subject to conservation covenants for protection. The high proportion of range on private land or state forest led Munks *et al.* (2004) to recommend (i) no conversion of potential habitat to plantation or clearing for agriculture within the range of either species, and (ii) a moratorium on clearfelling, burning and sowing silviculture in any potential habitat within the range of *H. bornemisszai* until their potential long-term effects have been assessed. More generally, but implicit in the approaches outlined above, Michaels and Bornemissza (1999) concluded that conservation of this lucanid fauna necessitates forest management at the landscape scale, with the central need to ensure supply of dead trees in conditions suitable for the entire diverse saproxylic fauna.

The 17 South African species of *Colophon* (see p. 120) are considered to be a relict montane group, and are the only stag beetles found at higher altitudes in the Western and Southern Cape, with the sites believed to be refuges now separated and effectively isolated by changes in land use and climate at lower altitudes (Geertsema 2004). As presumed cold-adapted species, *Colophon* may be susceptible to global warming, but some of their montane sites are threatened by establishment of repeater stations for mobile phone services, and the accompanying roads may open up some previously remote areas and facilitate access by unauthorized or commercial collectors (see p. 120).

Jewel beetles

Buprestidae have long been attractive to collectors. As their common name implies, many are brilliantly coloured and exotic in appearance, and many are also restricted in distribution. Their attractiveness has led to their use as decorative objects in many cultures, and to value as collectables. For Australia, for example, Buprestidae have been the most frequently advertised beetles by European insect dealers, with 59 species recorded as offered for sale in the previous decade (Hawkeswood *et al.* 1991). Most of these species are believed to have been captured in the wild rather than being captive-bred, and are endemic to parts of Western Australia, where Buprestidae are formally protected (see p. 38). In general, the larger and more colourful species command higher prices than smaller or less striking taxa. Several species have also been employed as biological control agents against pest weeds. For example, two European species have been imported to North America in this context: *Sphenoptera jugoslavica* (feeding on the rangeland weed *Centaurea diffusa*, diffuse knapweed) and *Agrilus hyperici* (on *Hypericum perforatum*, St John's wort) and are thus regarded as beneficial species.

Other species are recognized pests, and various buprestids attack either living or dead trees. For example, the emerald ash borer *Agrilus planipennis* is native to South-east Asia but is a major threat to native *Fraxinus* (ash) species in North America. Species of *Melanophila* are well known as being attracted to forest fires, sometimes from considerable distances, and depend on freshly fire-killed trees. Their larvae cannot overcome the defences of living trees. Indeed, the females may oviposit under bark as soon as the flames have subsided, and respond to infrared radiation (via specialized sensory receptors; Schmitz & Bleckmann 1998) to detect fires.

Ladybirds

Ladybirds (or ladybugs) are regarded widely as harbingers of good tidings, and are popular and well-liked beetles (Majerus 1994). Many are colourful and easily recognizable, enhancing their values as conspicuous invertebrates, particularly in parts of the northern hemisphere (Iperti 1999). The family Coccinellidae is not particularly large, containing around 4500 species, and the majority of species are predators, leading to considerable practical interest in their use as manipulable biological control agents of aphids and coccids (Dixon 2000), with considerable segregation between the species favouring one or other of these predominant prey groups. Some ladybirds are regarded as among the most important predators on crop pests such as aphids and other Homoptera. Indeed, several studies comparing Coccinellidae diversity in natural and managed habitats such as crop systems have found that richness and abundance is greater in crop systems than in more natural environments (Duelli 1988; Magagula & Samways 2001), with the richness determined in part by prey abundance. Assemblages of Coccinellidae may be considerably less rich than those of many other beetle families, perhaps reasonably so for relatively specialized predators. However, local richness may commonly be in the range of 10–25 species, sufficient to provide useful indications of change and dynamics. Many species tend to occur on a limited range of vegetation types.

Much of their conservation interest relates directly to their predatory role, and ladybirds have become embroiled in the wide-ranging and continuing debate about the non-target impacts of polyphagous predators introduced as classical biological control agents, with the possibility of such alien species causing harm to native taxa. For example, *Coccinella septempunctata* was considered a possible threat to the endangered Karner blue butterfly *Lycaeides melissa samuelis* (Lycaenidae), because of synchronization of ladybird larvae and adults with butterfly eggs and larvae (Schellhorn *et al.* 2005). Establishment of an isolation distance between cropping areas supporting aphid populations through which ladybird populations could increase and sites supporting Karner blue butterflies was considered a valuable component of the butterfly management programme.

Some such correlative studies link expansion of adventive coccinellids with declines in native ladybirds, but strong causative evidence is harder to obtain. The decline of *Coccinella novempunctata* (C9) in North America (Harmon *et al.* 2007) exemplifies the interpretative problems that can arise. Both C9 and the two-spotted ladybird *Adalia bipunctata* have become rare across their formerly wide North American range, with two adventive species, *Coccinella septempunctata* and *Harmonia axyridis*, considered likely causes of this decline and of changes in the composition of native ladybird assemblages. Harmon *et al.* noted that at least 179 coccinellid species have been introduced, either intentionally or unwittingly, into North America and that 27 of these have become established. *Coccinella septempunctata* and *H. axyridis* are among the most important adventive species found on agricultural crops, with the latter having recently become the more common species in many surveys. A third alien species, *Propylea quatuordecimpunctata*, has also become very common in eastern North America (Alyokhin & Sewell 2004). The mode of arrival of all these species in North America remains uncertain. Early efforts to introduce all of them deliberately as biological control agents are widely believed not to have succeeded, as they were not found in subsequent surveys at and near the release sites (Day *et al.* 1994). More recently, most sightings have been found around seaports, strengthening the belief that they might have been introduced accidently on ships. Alien ladybirds may sometimes take a considerable period to spread. The Holarctic *Adalia bipunctata* was first recorded in Japan in 1993, and for several years was known only from a 25-ha park in Osaka. Since then it has spread through the neighbouring region. Toda and Sakuratani (2006) noted that increasing prey and host-tree range have also increased, suggesting that the initially slow colonization pattern may be expected to accelerate in the future.

Harmonia axyridis has recently aroused considerable concerns over its invasibility and effects on non-target species, and is viewed widely as a serious pest species (Roy & Wajnberg 2008) (see p. 122). In addition to effects of direct conservation concern, it has in places become a pest of fruit, through the aggregations of adult beetles that may occur. In viticulture particularly, it is difficult to remove ladybirds sheltering among bunches of grapes so that the alkaloids from crushed *Harmonia* may spoil the resulting wine and result in commercial losses. Considerable background to this species was summarized by Koch (2003, see also Koch *et al.* 2006), who cited several cases in which this ladybird appeared to be the top predator in guilds of aphidophagous insects by using them as food. It thereby participates in reduction of native species abundance and changes in guild composition in orchards and field crops. The mechanisms of displacement of native coccinellids are perhaps complex, and involve relatively high voracity and fecundity, as well as polyphagy (which extends to feeding on members of other predatory groups, such as Neuroptera). However, the roles of most other introduced ladybird species in causing declines of native species often remain contentious, in common with other non-target effects (Obrycki & Kring 1998; Obrycki *et al.* 2000), and difficult to appraise. Such dilemmas arose initially following the use of the Australian *Rodolia cardinalis* against cottony cushion scale *Icerya purchasi* on *Citrus* in California in the late 1880s (Caltagirone & Doutt 1989). That case stimulated wide enthusiasm for use of ladybirds in biological control, and a massive number of what Caltagirone and Doutt termed 'haphazard introductions' resulted in many parts of the world. Many of these were not documented adequately, and successful introductions can often be inferred only by the presence of known adventive species as resident populations, with little if any knowledge of how long ago they were introduced for biological control or arrived by other means.

A feature of the life cycles of some ladybirds is their aggregations in vast numbers for hibernation, circumstances that have been suggested to increase the risk of mortality caused by mycosis due to *Beauveria* fungus. Some overwintering species have declined considerably. One of these, *Semidalia undecimnotata* in Europe, was appraised by Iperti (1999) with a view to suggesting potential remedial measures. Declines in overwintering adults reflect vulnerability to environmental changes in several distinct contexts: (i) increase in pollution; (ii) increased cultivation of cereal crops, reflecting that this ladybird's predominant prey (aphids such as *Aphis fabae*) are associated with leguminous crops; (iii) impacts of climate change resulting from industrial pollution in Europe; and (iv) installations of radioelectric instruments on mountain tops, known to kill large numbers of adults. Artificial shelters for hibernating *S. undecimnotata* have been placed in such sites to provide additional shelter and protection. Iperti suggested that the conservation needs for this species change during a year, initially to protect larvae present on crops but later to protect mountain summits to which adults migrate. However, much of the impetus from Iperti's suggestions was to maintain high densities on crops, as beneficial fauna, whereby modifications of pesticide regimes and some level of habitat manipulation may be beneficial.

Provision of refuges for crop coccinellids may, for example, necessitate cultural practices such as cutting grass in orchards and strip harvesting of some crops. Augmentation of field populations from insectary-reared stocks may also be needed.

Longhorn beetles

Longhorns, also termed longicorns or long-horned timber beetles and members of the family Cerambycidae, parallel Lucanidae in containing some of the most imposing of all beetles and also being timber borers; however, they differ from lucanids in that many are important pests of forest trees and may demand management to curtail their impacts on commercial forest crops. Indeed, Linsley (1959) described Cerambycidae as 'one of the economically most important groups of insects of the world', because of larval boring in trees and, more rarely, other plants. Much of the impetus for their biological study arises directly from this importance. Hanks (1999) recognized four broad categories of larval host plants: healthy hosts, weakened hosts, stressed hosts and dead hosts, the last commonly supporting several generations of these borers. Some cerambycids are attracted to recently fire-killed trees, with (for conifers) smoke and turpentines both among the attractant principles for *Monochamus*. In addition, even a dark silhouette may be attractive (de Groot & Nott 2001), and some species respond to the kairomones of pine bark beetles.

One important consequence of this habit is the relative ease with which the beetles can be introduced to new areas as cryptic larvae within lumber or wooden packing materials. Some may then become pests in their new range, bur others simply constitute records of aliens. Thus one of the two introduced species reported from the Galapagos (*Neosema kuscheli*, known from South America) is known there from only one specimen, believed to have been imported in this way (Peck 2006). The other Galapagos introduction (*Carphina arcifera*, from Mexico and Venezuela) exemplifies another possible mode of transport and entry, possibly as a stowaway initially attracted to a boat light.

The family also includes some of the world's largest beetles, including the South American *Titanus giganteus* which can sometimes exceed 20 cm in body length. The common names of the family reflect their most obvious structural feature, elongated antennae that range from around two-thirds to more than three times the body length. Many are also brightly coloured, and thereby attractive to collectors. In Europe, the striking *Rosalia alpina*, associated with deciduous beech forests, is recognized as of conservation interest in most of its range countries and is a notable flagship species for assemblages of saproxylic and other forest beetles. It is threatened by changes in forestry practice, particularly the removal of felled wood from forests, rather than leaving this resource *in situ* for substantial periods, and by the fact that plantation forests are not grown to stages sufficient for colonization. Restoration or reservation of key habitats, as for so many other beetles, is a key need. Similar considerations for a very different cerambycid, the Valley elderberry longhorn beetle *Desmocerus californicus dimorphus*, a riparian species found in association with blue elderberry

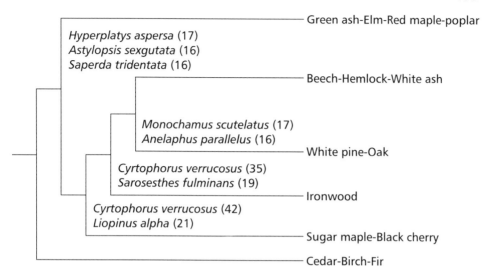

Fig. 9.5 Cerambycidae as indicators of forest diversity in North America: representation of species with significant indicator values for tree species site clusters. Figures in parentheses are IndVal values. (From Holland 2007 with permission.)

(*Sambucus mexicanus*) in California's Central Valley, led to establishment of conservation banks (Talley *et al.* 2006, 2007). The five such banks are centralized conservation areas (easements) in which projects that have impacts on the species' habitat may purchase entitlements (credits) to offset the effects of their projects, with the funds used for restoration of the areas.

Cerambycid richness in an area may be supported by a combination of coarse woody debris and flowers, the latter important food sources for adults. Several cerambycid species in North America are indicators of high diversity of tree species and large size (> 10 cm diameter at breast height) (Holland 2007). Across 190 sites, tree size and species composition largely determined the amount and character of potential habitat for larvae, so influencing the species able to develop there. Several species were weak but significant indicators of site tree groups (Fig. 9.5) (Holland 2007), whilst *Cyrtophorus verrucosus* had higher indicator values in two different tree associations: such breadth, of course, militates against it being an indicator in other contexts, for which specificity may be important.

One significant threat to longicorn richness, as for other forest beetles, is replacement of multispecies broadleaved forests by monoculture conifer plantations, as has occurred in many places. In central Japan, high abundance of cerambycids in young conifer stands was associated with floral diversity (Makino *et al.* 2007), although this result reversed that found in broadleaved forest stands, in which richness was lowest in the young stands (Sayama *et al.* 2005), possibly due to differing amounts of coarse woody debris and floral diversity. Differences in richness between plantations of larch and broadleaved forests (as found by Ohsawa 2004b) led Makino *et al.* to recommend that, wherever possible, stands of mixed broadleaved forest should be retained interspersed with conifer plantations.

However, species richnesss of longhorns was higher in old-growth forests than in second-growth forests of conifer plantations in south-western Japan (Maeto *et al.* 2002), with the greatest contrast between old-growth and conifers. Some taxa appear to need the conditions provided by old-growth forests, possibly related to trees/branches with large diameter. For example, six species of *Pidonia* occur under bark of a variety of large trees and branches, and seem to fall into this category. Extended rotation periods or delays in cutting are thereby conservation considerations. Thinning may be valuable, in enhancing growth rates of the remaining trees. Maeto *et al.* suggested that *Pidonia* might be monitored as indicators of the progress of forest restoration.

10

Concluding Thoughts

Many beetle taxa other than the families noted in the previous chapter have, of course, been signalled as of conservation interest, but many of their conservation needs have indeed been relatively slow to be acknowledged. For many years, the European stag beetle *Lucanus cervus* was the only beetle listed for national protection by several range countries, simply because it aroused concerns as an impressive, important and easily recognizable flagship species. And more modern listings in many parts of Europe are dominated by members of the 'popular' families such as Carabidae, Scarabaeidae, Lucanidae and Cerambycidae, for which documentation is generally sufficient to present compelling evidence of threat and to make convincing cases for conservation need. Rare species of weevils, leaf beetles, ladybirds and others also occur in such compilations but, in general, largely as the consequence of the zeal of individual proponents and not as components of any more comprehensive overview. Syntheses of beetle conservation requirements paralleling those available for the most well-known or popular taxonomic groups or selected ecological groups (such as saproxylic beetles, riparian carabids) are not available for most taxonomic or ecological groups. However, the available syntheses suggest both important ways forward and important limitations to this approach to conservation.

Individual species conservation plans or recovery plans for beetles (and others), even when the major needs are displayed comprehensively and clearly, are commonly difficult to bring to fruition. In many instances, however, those difficulties are exacerbated by lack of clear focus and objectives in the management plan. Documents prepared under direction from different authorities, probably with different terms of reference or obligation, and in different parts of the world can vary enormously in length and scope. They are likely to share the objectives of (i) minimizing or eliminating the risk of extinction in the short term and

Beetles in Conservation, 1st edition. By T.R. New. Published 2010 by Blackwell Publishing.

(ii) providing conditions for long-term sustainability, either in sites the beetle occupies at present or elsewhere. Plans range from short overview documents to very detailed management protocols, but each should encapsulate the reasons for conservation concern (threats) as a basis for remedial management, and set out the management objectives and actions needed for each, ideally using so-called SMART terms (see p. 60), so avoiding vague and open-ended statements in favour of well-focused and planned measures. These needs are not peculiar to beetle plans, of course, but may be particularly useful to potential managers who are not familiar with the focal taxa in that they help to provide unambiguous direction and focus and also indicate the partitioning of needs between practical management and more basic research to inform this. Without such direction, misunderstandings are more likely to occur, and research extend well beyond the questions of direct management relevance. Some projects have made little practical progress, despite generating valuable publicity and focus over several decades. The Valley elderberry longhorn (see p. 192) was listed in the USA in 1980, and the initial recovery objectives (United States Fish and Wildlife Service 1994) are as follows.

1   Preserve and protect known habitat sites to provide adequate conditions for the beetle.
2   Survey Central Valley rivers for remaining beetle colonies and habitat, and incorporate findings into short- and long-term management plans.
3   Provide protection to remaining beetle habitat within its suspected historic range.
4   Determine number of sites and populations necessary to eventually delist the species.

Yet Talley *et al.* (2006) noted that recovery criteria (i.e. 'Measurable' from SMART) for the longhorn had still to be defined, pending more information on its biology and ecology so that, despite lack of substantial progress, after more than 20 years the 1984 objectives still 'show the durability of the basic tenets of conservation biology – protect and restore habitat, address threats, reestablish populations'. For objectives 1 and 4, protective regulation occurred for most habitats but losses of known and potential habitat continued, together with resistance to conservation for fear of land-use restrictions on some private land. Surveys (objective 2) are still lacking for the southern parts of the range, and long-term surveys to examine trends with land-use changes are sparse. Talley *et al.* (2007) noted that collaborations are needed between the various stakeholders to effect habitat management plans for some important areas. As for objective 3, despite some progress 'dramatic gaps and gray areas remain in our understanding of the species' (Talley *et al.* 2006, p. 62), so that recovery criteria cannot yet be defined, and long-term data on population dynamics, intergenerational changes and patch turnover are not available (see Talley 2007).

Implications from this plan are much more widespread. Many such individual species plans for insects, in many places, convey strong initial goodwill and sound basic scenarios for making progress, but the goals become frustrated as their generality and the difficulties of advancing objectives in practice become apparent. A general problem, discussed by New (2009), is that many of the

managers on whom conservation management obligations devolve are not entomologists, and are commonly not familiar with insect biology. However well
intentioned, direct or uncritical transfer of experiences with vertebrate animals
or vascular plant management to beetles may not be wholly satisfactory, not
least because managers may be operating outside their competency in taking
important decisions on conservation activity. One outcome is that constructive
conservation progress and practical management may be irregular, sporadic and
insufficiently integrated as adaptive management for the beetle, and replacement
of general themes by more specific objectives and actions is relatively rare. More
precise objectives presuppose more background knowledge, but obtaining that
knowledge must be predicated on habitat (study site) security to enable study
to proceed and the means to undertake the requisite surveys or monitoring exercises, supported by the availability of specialists from whom constructive advice
may be solicited. Monitoring must occur at regular (often intergenerational)
intervals and be planned to provide targeted information beyond simple presence or absence. Many beetle management plans viewed in retrospect show missed
opportunities, reflecting the wider problems of establishing the programme and
assuring its continuity as personnel, funding demands, conflicting needs and
local priorities, and agency duties change. It is almost inevitable that levels of
interest, capability and commitment to any species recovery plan will change
over the decade or more needed to refine it and bring it to fruition. However,
the need to ensure that limited support for species conservation is used effectively demands that management plans should be tightly focused and adequately
resourced, and that the plans set out are accountable. The optimal balance between
research and management is often very difficult to determine, with the former
driven by key aspects of management need.

Most such species conservation plans have initially been restricted to particular
countries or regions within a country, but wider geographical perspective is a
welcome trend in aiding less parochial efforts for some species, promoting their
values as umbrella or flagship taxa for wider conservation isssues. For Europe, the
status of *Osmoderma eremita* (see p. 25) has been appraised for 33 countries
as an important flagship for the multitude of saproxylic invertebrates in the region
(Ranius *et al.* 2005). This survey revealed that *Osmoderma* seems to have declined
in all European range countries, but also confirmed its relatively wide host range
with the primary importance of oaks (*Quercus*). Steps towards a broad unified conservation plan for this species can incorporate conservation issues of much wider
significance for Europe: (i) preservation of remnants of natural forest with old
broadleaved trees; (ii) preservation of remaining patches of forests in urban areas;
and (iii) preservation and restoration of habitats restricted to traditional agricultural
landscapes. The presence of *O. eremita* in areas long used by people reflects continuing presence there of large old trees, and declines reflect the progressive loss
of these with changing management and factors such as the fear of human injury
from falling wood in urban parks. Collectively, the grouped conservation measures noted above contribute to many other conservation priorities in Europe.

Whereas the *Osmoderma* studies have largely been undertaken by professional
scientists, some more conspicuous beetles attract far greater public attention, so
that conservation measures can incorporate this wider interest and appreciation

effectively. The European stag beetle *Lucanus cervus* (see p. 138), also sapro-xylic, is one such example of an appealing flagship species that has declined over much its European range. Widely distributed documents in Britain and elsewhere emphasize the relatively simple principles fundamental in site management, in addition to helping collect and coordinate incidence records. The emphasis in public advice on *L. cervus* is on simple practical measures that can be under-taken or promoted easily by individuals or sympathetic local authorities (Frith 2000). Suggestions for site management include (i) retaining as much dead wood as possible on site; (ii) leaving windblown trees in place unless they pose safety problems; (iii) leaving soils and vegetation around dead wood undisturbed, and ensuring that as much dead wood as possible is on or near the ground; (iv) avoiding stump grinding where possible; and (v) providing for measures such as these to be taken in wider site management plans where the stag beetle is known or suspected to occur. Creation of 'loggeries' (Fig. 10.1) in shaded positions in gardens within the beetle's range may provide refuges for beetles, and other simple domestic conservation measures may include rescue of beetles fallen into garden ponds, and protecting beetles from predators such as cats.

However much 'value adding' can occur for any individual species, not all threatened beetles can be treated case by case. The vast numbers of beetles of conservation interest effectively demonstrate the logistic limits of the individual species level as the main conservation approach: our resources and expertise are simply insufficient, and triage systems may be unconvincing. Despite the enor-mous political and communication importance of selecting a variety of deserving species for individual treatment, the approaches of conjoint plans emphasizing the complementary needs of species co-occurring in specialized biotopes, or species that are otherwise ecologically linked, are likely to increase in importance and relevance as being seen to have higher benefits for the conservation efforts involved.

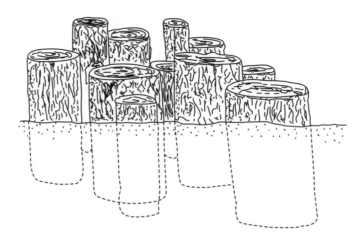

Fig. 10.1 General construction of an artificial loggery as habitat for the stag beetle *Lucanus cervus* in Britain. Hardwood logs (10–50 cm diameter) with bark attached are sunk about 60 cm into the ground in partially shaded areas. (After Frith 2000 with permission.)

The three grouped plans for British beetles noted earlier (see p. 37) exemplify some aspects of this approach. The first of these includes saproxylic beetles, with the 10 focal species all associated with veteran trees in old deciduous woodland: five of them are known in Britain only from the Windsor area, and the 10 comprise six species of Elateridae and single members of Scarabaeidae, Dryophthoridae, Eucnemidae and Melyridae. Most are classified as endangered in Britain, and all are rare or declining in Europe. Concern over declines relates to land-use changes involving loss of old woodlands and deterioration of old veteran trees, together with wider removals of dead wood and loss of temporal continuity across native woodlands. The major objective of the statement is to 'maintain the range of each species', with proposed actions restricted to monitoring and searching, with the practical conservation needs likely to be catered for by action plans for other species (such as *Lucanus cervus* and the violet click beetle *Limoniscus violaceus*, also very rare throughout its European range, dependent on old trees and listed under Appendix 2 of the European Union Habitats Directive) in the same habitats in the UK. These thereby act as umbrellas for the less conspicuous saproxylic organisms, but the latter are present as individually cited species on the formal conservation agenda rather than simply being part of the unknown invertebrate complement more usual in such contexts.

The second UK plan covers two species of *Harpalus* found on coastal dune grassland and inland chalk grassland, *H. cordatus* and *H. parallelus*. Both species are classified as rare, and threatened by dune stabilization and recreational use of coastal dunes, as well as more widespread inappropriate management of chalk grasslands. Each site involved is designated as a Site of Special Scientific Interest and the plan objective is to maintain populations at all sites. Again, the principal activity is monitoring and the need to consider the requirements of the beetles in the delivery of action plans for other taxa in these biotopes. The third group plan covers six species of river shingle beetles (three Carabidae, two Staphylinidae and one hydrophilid), associated by living on exposed riverine shingle areas, and subject to declines caused by many forms of land-use change with impacts on these restricted habitats. Exposed riverine sediments, of which shingle is a notable category, are a feature of current conservation concern in Britain (see p. 80), and the aim of maintaining viable populations of these beetles within each occupied catchment is augmented by one to enhance populations at selected sites. A variety of policy and more active practical measures are included, each with designated lead agencies suggested for bringing them to fruition. These three plans demonstrate an important umbrella effect, whereby beetle species are noted as components of a portfolio of need, within a strong taxonomic framework that enables notable species to be recognized and acknowledged in wider conservation plans.

Umbrella messages are easier to promote if the sheltering species is a notable vertebrate or popular plant, but such a species as an umbrella for sites with documented threatened insects is far different from the widely advocated approach that beetles (and other invertebrates) will be conserved automatically as unknown passengers without specific attention as long as vertebrates present in the same area are conserved. In Finland and Russia, the white-backed woodpecker *Dendrocopas leucotos* is endangered, and surveys of 16 potential forest sites for the bird yielded 16 threatened beetle species and 23 uncommon species (in a pool of 300 species collected), with each site inferred to include six or more threatened species (Martikainen *et al.* 1998). Most of the significant species depend on dead wood and, with the woodpecker requiring areas of about 50–100 ha of habitat to thrive, there is a strong likelihood that viable populations of the beetles could be sustained in the same areas, so that a key to conserving some threatened saproxylic beetles may indeed be to conserve the woodpecker and the dead wood it needs. However, it is still unclear whether some of the beetles are even more sensitive than the woodpecker to habitat degradation (Martikainen *et al.* 1998), so that even losses of small numbers of trees in forests or increased stand isolation may be significant. There is also some chance that the woodpecker itself might become extinct in Finland, in which case the umbrella justification disappears and the beetles revert to needing conservation in their own right and without vertebrate support. Continuing to accrue knowedge of the specific needy beetles is thereby a sound investment in their future con-servation, even if this is not a priority at present.

For much of the world, this level of detail is impossible to achieve. Whereas patches of tropical forest may be selected for active conservation on the presence of notable birds or mammals, only very rarely is there any detailed knowledge of what beetles may be present and very little on their conservation status or needs, other than by inference. In those environments, despite massively increased understanding over the last few decades, even the most complete available data on richness and biology of the best-known beetle groups is sadly inadequate to formulate specific management. Thus, arboreal Carabidae in a small area of Ecuadorian Amazonia sampled by fumigation comprised 318 species, of which more than half were undescribed (Lucky *et al.* 2002). Moreover, there was very high turnover of species composition between sampling sites with, in general, fewer than half the species shared between most sampling dates, so that species composition changed substantially across seasons. This suggested that the total species pool may be very large. Only 10 species were represented by more than 50 individuals each (collectively comprising 31.9% of individuals) over the study which extended for 3 years and included 900 canopy fogging samples across 100 sites; 91 species were represented by singletons.

This study is large and, as Lucky *et al.* (2002) commented, Carabidae is a reasonably well-studied family so that the discovery of so many previously unknown species indicates the extremely incomplete knowledge of tropical bee-tles. A related interpretative problem is that the reasons for the apparent rarity of so many species in such surveys are also speculative. It is often not clear whether an impression of rarity results from inadequate sampling approaches or genuine scarcity. The examples of constitution of tropical beetle faunas noted in Chapter 1 (see p. 10) could be multiplied, but collectively demonstrate the dif-ficulties in understanding against which practical conservation must be driven.

In common with approaches to diverse beetle faunas with numerous poorly documented species in temperate regions (Romero-Alcaraz & Avila 2000 for Spain), conservation tactics are then likely to devolve on preserving environmental heterogeneity and as much variety as possible, emphasizing the importance of more natural areas and the need to protect them from despoliation. Wherever they may be, every site is likely to differ from every other one. Even tentative interpretations of possible reasons for such differences (vegetation, cultivation intensity, disturbance, possible gradients) may provide valuable clues to management, with the major categories being (i) extent and forms of human activity and change; (ii) abiotic gradients (altitude, topography, orientation and slope, soil features); and (iii) biotic factors (vegetation, successional stages, overall diversity, interactions with other taxa). Landscape heterogeneity, both of form and use, may be key determinants of local beetle diversity and well-being. Site reservation or other effective protection is an important initial step, but may be very difficult to achieve, and subsequently to monitor.

Few conservationists doubt this central need to conserve natural habitats and to slow anthropogenic changes, and to promote the connectivity of potentially isolated sites within the landscape (by development of habitat networks; Samways 2007) and buffering their transitions to agroecosystems or other developed regions. Whether focusing on small patches of forest or heathland in temperate regions or on tropical forests or savannas, practicality may be very difficult, but perhaps nowhere more so than in densely populated tropical regions where human needs are severe and knowledge of the biota is fragmentary. As Grove and Stork (1999) put it, 'The sheer diversity of forest insects worldwide presents an enormous challenge to insect conservationists grappling with the major issues in forestry and conservation today'. Almost any such area, however small, will yield high beetle richness with even moderate sampling effort, and the samples are likely to include unusual or rare species in numbers sufficient to evince expressions of conservation need.

Many of the problems of studying largely unknown insect diversity in the tropics were discussed by Lewis and Basset (2007), but knowing the problems is only a first step; gaining that knowedge does not necessarily lead to solving them. Theory and ideas, however convincing they may be, do not translate easily to effective conservation in such environments. Nevertheless, any research agenda seeking to explore ways of increasing harmony between land use and conservation may benefit from including information on beetle biology. The substantial information on boreal and temperate forest saproxylic beetles, for example, implies strongly that continuity depends on the critical resource of dead wood, that many species have become threatened by logging and sanitation associated with forestry, that even small forest patches are important as refuges, and that the group as a whole may be functionally significant in forest ecosystems. This framework contributed to a suggested research agenda on saproxylic insects and logging impacts in tropical forests, for which Grove and Stork (1999) set out their key questions as follows.

1 Can suitable techniques be found to effectively sample and enumerate these insects in tropical forests, and can such an approach be used widely?
2 Do logged forests actually differ in saproxylic insect assemblages from old-growth forests, so is their conservation really an issue in the tropics?
3 If logging effects occur, are they sufficiently short term (i.e. of a decade or so) to be detected within a reasonable or practicable period, or do they take up to several logging cycles to manifest?

4 If logging is sufficiently non-intensive to enable survival of more characteristic megafauna, does this also effectively conserve saproxylic insects or is
   this possible correlation spurious?
5 Can researchers determine the key dead-wood habitats and say which habitats
   and species may be under particular threat from logging?
6 Can forestry growth/yield models assess long-term supply of overmature trees
   and dead-wood resources under a range of silvicultural conditions?
7 Is there any simple correlation between saproxylic community integrity and
   the processes of wood decomposition and nutrient recycling, and is this affected
   by logging influences on the insect communities?

Similar schemes could be designed for other disturbance processes, in a variety
of biotopes and regions, and beetles might become an important focal taxon in
these. In such contexts, despite lack of detailed species-level taxonomy, the morphospecies approach can often (at least tentatively) be associated with sensible
presumptions of trophic guild allocations. In the above example, this means
that various categories of saprophages may be inferred strongly to be functional
groups in aiding ecological interpretation. Despite the need for caution in any
extrapolations of this nature, Grove and Stork (1999) suggested that a few
autecological studies of candidate tropical taxa could help to clarify this key
issue of understanding.

The disturbance/productivity approach advanced by Eyre (2006) (see p. 70)
may have potential for some selected beetle assessments in the tropics, but the
main stumbling block remains lack of capability to accumulate sufficient data
to ensure robust interpretations. The same holds true for much of the southern
temperate region. For much of the world, beetle conservation, in common with
that of most other terrestrial and freshwater invertebrates, must depend very
largely on effective protection of key biotopes from despoliation, as reserves
within which human intrusions are restricted and monitored. The difficulties
of achieving this, even in places traditionally considered to be remote (such as
parts of New Guinea), are formidable and even within acknowledged global
biodiversity hotspot areas, the urgent effective protection of natural areas compatible with human needs is often not accorded high local priority. Quite
understandably, human needs take precedence. Systematic attempts at biological
inventory (of beetles or any other major taxonomic group) and long-term
monitoring (of a decade or more) generally do not exist. Also generally, many
remaining 'natural' habitats treated as important conservation refuges or remnants
are by no means pristine, many are small, and most are under some form of
human pressure. Habitat protection is the paramount theme in insect conservation and whilst many entomologists acknowledge readily that securing habitat
(place) does not ensure conservation of all species present there, it is even more
evident that without a place to live those species will not survive. Habitat (place)
security is simply the foundation for current management of resources needed
by the occupants. Vegetation succession, climate change and other factors may
in course render the place unsuitable. Almost invariably, practical considerations
for beetle conservation address only the present or short-term future priorities,
as basic need for possible more adaptive management in the future as such

changes eventuate. Protected areas could provide much valuable information on environmental changes should periodic surveys contributing to inventory be instituted as an investment in understanding, with beetles a likely useful candidate group for inclusion. As noted earlier (see p. 31), there are few such studies in or near the tropics. Howden and Howden (2001) examined intermittent samples of dung beetles from a wildlife refuge in southern Texas, and showed that only 40% of the species persisted there over a 30-year period between first and last surveys, and that the relative abundance of native species declined over that period.

Because of their richness and ecological variety, beetles could serve as a major facilitator informing conservation decisions affecting the biota and environments of virtually all major terrestrial and freshwater environments. However, as noted above, the rules of thumb needed to affirm this are uncertain. There is wide consensus that inventory, consistent recognition of the constituents at the species (or near-species) level and of their functional roles, and their responses to change, together with possible surrogate values for parallels in other taxonomic groups and ecosystems, are desirable. However, application of this information in the context of practical conservation, essentially assessing the tolerable amounts of anthropogenic disturbance and their consequences, is also difficult. Much of the species approach to beetle conservation, because it emphasizes restricted range and ecological specialist species, elicits responses to queries about intrusion tolerance of 'none', 'very little' or 'very carefully planned and regulated'; wider assemblage focus is likely to shift this response to greater tolerance if the majority of core species are not as sensitive to small changes. As they increase in number and diversity, individual species considerations inevitably lead to greater land variety and more of the landscape being deemed desirable, and even necessary, for conservation. Even large enclosures or reserves (as above) may not be effective. Management for any single species, particularly on small areas, may compromise the existence of other species, even rendering them increasingly vulnerable if the balance of resources is shifted uncritically. Practical emphasis then shifts from the ethical ideal of conserving every species present to attempts to determine the effects of predictable non-stochastic disturbance regimes on the richness and representativeness of assemblages. Lewis and Basset's (2007) 'wish list' for studies of disturbance effects on tropical insects is indeed relevant to beetles (Table 10.1), but much of this may long remain impracticable. Whatever a conservationist's ideals may be, human needs will continue to take precedence, but the major impediments to furthering conservation interest include lack of effective communication and of logistic support from within the areas of interest.

Using tiger beetles to illustrate their thesis, Pearson and Cassola (2007) demonstrated their increasingly focused role in conservation biology, in which descriptive natural history and basic species descriptions gradually give way to more sophisticated levels of study, with the inevitable consequence that much of the early audience or readership becomes effectively excluded from participation and understanding. Much of this divorce indeed reflects ineffective communication and failures by scientists to inform people other than their peers. The appraisal for tiger beetles, with comments such as 'rapidly growing use of

Table 10.1 'Wish list' for studies of the effects of disturbance on tropical insects.

Take into account the geographical distribution/endemicity of taxa, rather than focusing solely on overall species richness or diversity values

Report both species richness and diversity measures, and control for the critical influence of sample size on species richness through rarefaction

Be explicit about the nature of replication in the investigation

Document clearly the forms of habitat disturbance, and the time since disturbance events

Document the history of human and natural disturbance, and the time since disturbance events

To avoid publication bias, publish negative results (where no significant disturbance effect is found), as well as positive ones; this plea is addressed to editors as well as authors

Consider employing or exploiting experimental protocols, such as before–after control impact (BACI)

Use sound concepts of taxonomy (where morphospecies correspond to unnamed species) rather than fuzzy groupings of unidentified specimens

Use a multitaxon approach to reach more general conclusions as to the impacts of disturbance on diversity; where this is not possible recognize clearly the limitations associated with individual study taxa

Include summary data on numbers or individuals of each species recorded from individual sampling locations (perhaps as electronic appendices) to facilitate subsequent meta-analyses.

Source: Lewis & Basset (2007) with permission.

highly sophisticated disciplines such as molecular biology, statistical modeling and satellite imagery have introduced many technical words and concepts' and 'this trend includes increasing length and number of published articles, increasing sentence complexity, use of multi-word noun phrases, as well as narrowly-defined technical terms' (both quotations from Pearson & Cassola 2007, p. 53) conveys some idea of syndromes familiar to many conservation managers, whose primary need is for clear, unambiguous, non-technical and direct information in formats they can understand and use.

Much of the tropics, in particular, harbours few resident entomologists not concerned primarily with management of key pest species, and most conservation managers, anywhere, are not entomologists. Most of the information on tropical beetles noted in this book has come from the studies of visiting scientists, some of whom have made strenuous efforts to communicate their results to colleagues and managers in highly intelligible forms. All too often in history, though, such constructive feedback has not happened, and much information and the reference material (such as collections of specimens) on which it is founded is elsewhere. Many past collecting trips, scientific expeditions and similar exercises may be regarded as little more than smash-and-grab raids, with the loot (specimens, information) never being repatriated effectively. Despite much recent effort at more effective communication, much of the earlier colonial history of beetle collecting in the tropics was basically one of piracy, a time thankfully now largely past (with the notable exception of some overzealous commercial collecting; see Chapter 4). Technological advances, such as web-based identification systems based on excellent digital imagery easily and rapidly transmitted electronically, have already done much, and have far greater potential, to help identify specimens by reference to type material housed in far distant institutions. More accurate recognition of beetles of conservation interest is thus becoming feasible, as long as the basic computer infrastructure is available locally: it may not be.

One important limitation for conservation in the tropics is that we are already committed to working with and saving only a small portion of what existed a century or less ago. Janzen (1997), in an essay that challenges much of the externally imposed thought on conservation needs in the tropics, claimed that there are no unambiguously conserved tropical wildlands on good agricultural soils. If this is so, the current system of remnants of natural vegetation, whatever condition they are in, represents the predominant milieu for conservation action within the wider landscape, with augmentation from increasing hospitality across the intervening matrix (such as by changing agriculture or forestry management practices) and attempts to increase connectivity. Importantly, Janzen emphasized that for tropical conservation we do not really need much of the information treated as important by many temperate-region conservationists: we do not need to know, for example, exactly how many beetle species frequent this or that patch of forest, because we already know that there are a lot; knowing more would not contribute greatly to ensuring their security, and the substantial costs of refining that estimate could be better used for practical steps to aid that security. Focusing optimally into some of the agendas noted above may need very careful consideration.

At present, we have two major sources of information on beetles: the vast array of specimens (including those of fossil and subfossil taxa) accumulated in collections over some two centuries of endeavour and interest, and the beetles living in Earth's environments. Both need much more study but the scientific transition noted by Pearson and Cassola (2007), of natural history study giving way to more specialized and technical abilities, goes well beyond the institutional level to influence the development of individuals as entomologists or, in this context, as coleopterists or conservation biologists. Many of the leading beetle experts, and not a few other biologists, developed their interests through childhood collecting or curiosity about natural history, with those interests persisting and remaining influential. However, even in schools, simple observational natural history has become unfashionable and, importantly, difficult to teach because of safety issues, costs, reluctance to organize or participate in fieldwork and the practical difficulties of field trips, and the influence of the computer as a competing educational tool that provides virtual fieldwork without all these impediments. Likewise, we are increasingly seeing cohorts of university graduates with excellent technical skills but very limited practical experience with, or knowledge of, animals or plants in their natural surroundings. Some of the reasons for this were appraised by Cheesman and Key (2007), in a wide expression of concern for the future of entomology and the difficulties of recruiting young people into such important spheres of interest.

Many of the examples in this book have drawn on values of sound education and provision of information to the community (tourists and bark beetles,
see p. 3) and community involvement (volunteer recording of stag beetles, see
p. 39). Every case of beetle conservation or survey has potential to help gain
wider support for insect conservation activities. It is fair to claim that many insect
species management exercises could not proceed effectively without the help
of supportive volunteers, in addition to the non-scientist accumulation of historical information on distribution and biology. In much of the world, the
enthusiast volunteer helpers who participate in some exercises in the northern
hemisphere are simply not available. Fostering that interest and involvement,
for example by inclusion of local community stakeholders (such as affected
landholders, education bodies, environmental groups) on recovery teams, may
pay dividends and help to allay accusations of frivolity over 'saving a beetle'.

The richness and ubiquity of beetles render them eminently suitable as educational tools. Several recent invertebrate conservation programmes in Australia
have included specially prepared education kits for distribution and use in
primary schools and upwards. These are an important way of reaching young
people and eliciting wider interest and support. Enthusiastic teachers willing and
able to include lessons about beetles (or other insects) in their curricula, keep some
alive in the classroom, or undertake field trips (with due regard to any permits
needed for access or collecting) or simple outside exercises with their classes
have substantial potential to recruit interest for conservation, well beyond that of
many professional conservation managers. However, liaison between teachers
and managers can prove particularly beneficial, and similar partnerships also merit
serious consideration in more advanced, tertiary, studies. In conjunction with
this, a management team for any species may be able to suggest and encourage
such activities in different community sectors and also foster other collaborations,
not least through production of newsletters and other sources of information
for wide distribution, and being seen not to be remote from the local constituency.

For much of the future development of beetle conservation (as for many groups
of invertebrates) a prime need is the interest and commmitment of people who
are comfortable with studying the organisms in the field and able to appreciate
their systematics and diversity. In parallel with focused competent fieldwork,
interest in systematics is a primary need in documenting biodiversity and characterizing the entities involved. Many commentators have lamented the decline
of taxonomists in the biologist workforce (Lee 2000; New 2000), with many
of the most able institutional taxonomists not being replaced as they retire and,
as Cheesman and Key (2007) noted, severe problems in recruiting the next
generation of entomologists, including taxonomists. There is no easy solution,
other than through massively increased funding to redress this and to accord
such basic science greater appreciation. However, despite widespread calls for
fundamental documentation of biodiversity and for the means to achieve this,
governments and their major science funding agencies simply lack the cash and
commonly prefer to direct their limited support to more vote-worthy projects
further along the technological gradient than to basic descriptive taxonomy. Within
Coleoptera, the limited taxonomic support that does occur for conservation
might be usefully directed towards a carefully selected suite of 'catch-up' groups

(New 1999) to increase further the already broad relevance and applications of beetles in environmental assessments.

The current shortage of taxonomic expertise seems likely to guarantee that the bulk of the specimens in museum collections will reman orphaned (see p. 11) for the foreseeable future. In short, studying and sustaining Coleoptera requires coleopterists, and linking museum collections with field ecology requires broad perspective. Conservationists and biologists with other specialities may well be able to sustain the wider systems within which beetles live, but understanding the finer nuances of interaction between biologically varied and complex species and their resources so well demonstrated by many beetles may do much to enhance those efforts. Perhaps the central need is to seek a return to respectability for the study of natural history and, at least to some extent, to reverse the trend noted by Pearson and Cassola (2007) (see p. 203) in order to embrace those who like beetles and can communicate simply and clearly about them. Most of the foundation for beetle conservation has been in the observations and understanding of how beetles work, by people who progressively appreciate both the insects and the environments in which they live. Perhaps no conservation biologist would suffer, and most would have their capability and appreciations enhanced substantially, by even a superficial acquaintance with these powerful symbols of the wealth of invertebrate life.

References

Alaruikka, D., Kotze, D.J., Matveinen, K. & Niemela, J. (2002) Carabid beetle and spider assemblages along a forested urban-rural gradient in southern Finland. *Journal of Insect Conservation* 6, 195–206.

Alexander, K.N. (1999) Should dead wood be left in the sun or shade? *British Wildlife* 10, 342.

Alexander, K. (2003) Changing distributions of Cantharidae and Buprestidae within Great Britain (Coleoptera). In: *Proceedings of the 13th International Colloquium of the European Invertebrate Survey* (eds M. Reemer, P.J. van Helsdingen & R.M.J.C. Kleukes), pp. 87–91. European Invertebrate Survey, Leiden.

Alinvi, O., Ball, J.P., Danell, K., Hjalten, J. & Pettersson, R.B. (2007) Sampling saproxylic beetle assemblages in dead wood logs: comparing window and eclector traps to traditional bark sieving and a refinement. *Journal of Insect Conservation* 11, 99–112.

Allgower, K. (1980) Effect of the scarab beetle *Trox suberosus* on the hatching success of the east pacific green turtle *Chelonia mydas agassizi* in the Galapagos islands. *Annual Report of the Charles Darwin Research Station 1979*, 150–152.

Allison, A., Samuelson, G.A. & Miller, S.E. (1997) Patterns of beetle species diversity in *Castanopsis acuminatissima* (Fagaceae) trees studied with canopy fogging in mid montane New Guinea rainforest. In: *Canopy Arthropods* (eds N.E. Stork, J. Adis & R.K. Didham), pp. 224–236. Chapman & Hall, London.

Allsopp, P.G. (1997) Probability of describing an Australian scarab beetle: influence of body size and distribution. *Journal of Biogeography* 24, 717–724.

Allsopp, P.G. (1999) How localized are the distributions of Australian scarabs (Coleoptera, Scarabaeoidea)? *Diversity and Distributions* 5, 143–149.

Alyokhin, A. & Sewell, G. (2004) Changes in a lady beetle community following the establishment of three alien species. *Biological Invasions* 6, 463–471.

Andersen, A.N. (1999) My bioindicator or yours? Making the selection. *Journal of Insect Conservation* 3, 61–64.

Andersen, A.N., Braithwaite, R.W., Cook, G.D. *et al.* (1998) Fire research for conservation management in tropical savannas: introducing the Kapalga fire experiment. *Australian Journal of Ecology* 23, 95–110.

Andersen, J. & Hanssen, O. (2005) Riparian beetles, a unique, but vulnerable element in the fauna of Fennoscandia. *Biodiversity and Conservation* 14, 3497–3524.

Anderson, J.M. & Coe, M.J. (1974) Decomposition of elephant dung in an arid tropical environment. *Oecologia* 14, 111–125.

Andresen, E. (2003) Effect of forest fragmentation on dung beetle communities and functional consequences for plant regeneration. *Ecography* 26, 87–97.

Araya, K. (1993) Relationships between the decay types of dead wood and occurrences of lucanid beetles (Coleoptera: Lucanidae). *Applied Entomology and Zoology* **28**, 27–33.

Arndt, E., Aydin, N. & Aydin, G. (2005) Tourism impairs tiger beetle (Cicindelidae) populations: a case study in a Mediterranean beach habitat. *Journal of Insect Conservation* **9**, 201–206.

Arrow, G.J. (ed. Hincks, W.D.) (1951) *Horned Beetles. A Study of the Fantastic in Nature*. W. Junk, The Hague.

As, S. (1993) Are habitat islands islands? Woodliving beetles (Coleoptera) in deciduous forest fragments in boreal forest. *Ecography* **16**, 219–228.

Asher, J., Warren, M., Fox, R., Harding, P., Jeffcoate, G. & Jeffcoate, S. (2001) *The Millennium Atlas of Butterflies in Britain and Ireland*. Oxford University Press, Oxford.

Ashworth, A.C. (1996) The response of arctic Carabidae (Coleoptera) to climate change based on the fossil record of the Quaternary Period. *Annales Zoologicae Fennici* **33**, 125–131.

Ashworth, A.C. (2001) Perspectives on beetles and climate change. In: *Geological Perspectives of Global Climate Change* (eds L.C. Gerhard, W.E. Harrison & B.M. Hanson), pp. 153–168. American Association of Petroleum Geologists Studies in Geology 47. American Association of Petroleum Geologists, Tulsa, OK.

Assmann, T. & Janssen, J. (1999) The effects of habitat changes on the endangered ground beetle *Carabus nitens* (Coleoptera: Carabidae). *Journal of Insect Conservation* **3**, 107–116.

Asteraki, E.J., Hanks, C.B. & Clements, R.O. (1995) The influence of different types of grassland field margin on carabid beetle (Coleoptera, Carabidae) communities. *Agriculture, Ecosystems and Environment* **54**, 195–202.

Baehr, M. (1990) The carabid community living under bark of Australian eucalypts. In: *The Role of Ground Beetles in Ecological and Environmental Studies* (ed. N.E. Stork), pp. 3–11. Intercept, Andover.

Baker, S.C., Richardson, A.M.M., Barmuta, L.A. & Thomson, R. (2006) Why conservation reserves should not always be concentrated in riparian areas: a study of ground-dwelling beetles in wet eucalypt forest. *Biological Conservation* **133**, 156–168.

Balfour-Browne, F. (1962) *Water Beetles and Other Things: Half a Century's Work*. Blacklock Farries, Dumfries.

Balke, M., Kovac, D., Hendrich, L. & Flechtner, G. (2000) Rediscovery of the New Zealand diving beetle *Rhantus plantaris* Sharp, and notes on the south west Australian *R. simulans* Regimbart, with an identification key (Coleoptera: Dytiscidae). *New Zealand Journal of Zoology* **27**, 223–227.

Ball, G.E. (ed.) (1985) *Taxonomy, Phylogeny and Zoogeography of Beetles and Ants*. W. Junk, Dordrecht.

Banks, J.E. (2000) Natural vegetation on agroecosystems: pattern and scale of heterogeneity. In: *The Agroecology of Carabid Beetles* (ed. J.M. Holland), pp. 215–229. Intercept, Andover.

Barbero, E., Palestrini, C. & Rolando, A. (1999) Dung beetle conservation: effects of habitat and resource selection (Coleoptera: Scarabaeoidea). *Journal of Insect Conservation* **3**, 75–84.

Barr, T.C. (1969) Evolution of the Carabidae (Coleoptera) in the southern Appalachians. In: *The Distributional History of the Biota of the Southern Appalachians* (ed. P.C. Holt), pp. 67–92. Virginia Polytechnic Institute, Blacksburg, VA.

Barr, T.C. (1985) Pattern and process in speciation of trechine beetles in eastern North America (Coleoptera: Carabidae: Trechinae). In: *Taxonomy, Phylogeny and Zoogeography of Beetles and Ants* (ed. G.E. Ball), pp. 350–407. W. Junk, Dordrecht.

Barr, T.C. & Holsinger, J.R. (1985) Speciation in cave faunas. *Annual Review of Ecology and Systematics* **16**, 313–337.

Barratt, B.I.P. (2007) Conservation status of *Prodontria* (Coleoptera: Scarabaeidae) species in New Zealand. *Journal of Insect Conservation* **11**, 19–27. [Erratum **11**, 29–30.]

Bartlett, A.C. (1985) Guidelines for genetic diversity in laboratory colony establishment and maintenance. In: *Handbook of Insect Rearing* (eds P. Singh & R.F. Moore), vol. 1, pp. 7–17. Elsevier, Amsterdam.

Bartlett, R., Pickering, J., Gauld, I. & Windsor, D. (1999) Estimating global biodiversity: tropical beetles and wasps send different signals. *Ecological Entomology* **24**, 118–121.

Batary, P., Baldi, A., Szel, G., Podlussany, A., Rozner, I. & Erdos, S. (2007) Responses of grassland specialist and generalist beetles to management and landscape complexity. *Diversity and Distributions* **13**, 196–202.

Bates, A.J., Sadler, J.P. & Fowles, A.P. (2006) Condition-dependent dispersal of a patchily distributed riparian ground beetle in response to disturbance. *Oecologia* **150**, 50–60.

Bates, A.J., Sadler, J.P. & Fowles, A.P. (2007) Livestock trampling reduces the conservation value of beetle communities on high quality exposed riverine sediments. *Biodiversity and Conservation* 16, 1491–1509.

Baur, B., Zschokke, S., Coray, A., Schlapfer, A. & Erhardt, A. (2002) Habitat characteristics of the endangered flightless beetle *Dorcadion fuliginator* (Coleoptera: Cerambycidae): implications for conservation. *Biological Conservation* 105, 133–142.

Baur, B., Coray, A., Minoretti, N. & Zschokke, S. (2005) Dispersal of the endangered flightless beetle *Dorcadion fuliginator* (Coleoptera: Cerambycidae) in spatially restricted landscapes. *Biological Conservation* 124, 49–62.

Beaudoin-Ollivier, L., Bonaccorso, F., Aloysius, M. & Kasiki, M. (2003) Flight movement of *Scapanes australis australis* (Boisduval) (Coleoptera: Scarabaeidae: Dynastinae) in Papaua New Guinea: a radiotelemetry study. *Australian Journal of Entomology* 42, 367–372.

Bedick, J.C., Ratcliffe, B.C., Hoback, W.W. & Higley, L.G. (1999) Distribution, ecology, and population dynamics of the American burying beetle [*Nicrophorus americanus* Olivier (Coleoptera, Silphidae)] in south-central Nebraska, USA. *Journal of Insect Conservation* 3, 171–181.

Beebee, T.J.C. (2007) Population structure and its implications for conservation of the great silver beetle *Hydrophilus piceus* in Britain. *Freshwater Biology* 52, 2101–2111.

Bell, R.T. (1979) Zoogeography of Rhysodini: do beetles travel on driftwood? In: *Carabid Beetles: Their Evolution, Natural History, and Classification* (eds T.L. Erwin, G.E. Ball & D.R. Whitehead), pp. 331–342. W. Junk, The Hague.

Bell, R.T. (1985) Zoogeography and ecology of New Guinea Rhysodini (Coleoptera: Carabidae). In: *Taxonomy, Phylogeny and Zoogeography of Beetles and Ants* (ed. G.E. Ball), pp. 221–236. W. Junk, Dordrecht.

Berndt, L.A., Brockerhoff, E.G. & Jactel, H. (2008) Relevance of exotic pine plantations as a surrogate habitat for ground beetles (Carabidae) where native forest is rare. *Biodiversity and Conservation* 17, 1171–1185.

Biaggini, M., Consorti, R., Dapporto, L., Dellacasa, M., Paggetti, E. & Corti, C. (2007) The taxonomic level order as a possible tool for rapid assessment of arthropod diversity in agricultural landscapes. *Agriculture, Ecosystems and Environment* 122, 183–191.

Blanche, K.R., Andersen, A.N. & Ludwig, J.A. (2001) Rainfall-contingent detection of fire impacts: responses of beetles to experimental fire regimes. *Ecological Applications* 11, 86–96.

Bohac, J. (1990) Numerical estimation of the impact of terrestrial ecosystem by using the staphylinid communities. *Agrochemical Soil Science* 39, 565–568.

Bohac, J. (1999) Staphylinid beetles as indicators. *Agriculture, Ecosystems and Environment* 74, 357–372.

Bommarco, R. & Ekbom, B. (2000) Landscape management and resident generalist predators in annual crop systems. In: *Interchange of Insects between Agricultural and Surrounding Habitats* (eds B. Ekbom, M. Irwin & Y. Robert), pp. 169–182. Kluwer, Dordrecht.

Bommarco, R. & Fagan, W.F. (2002) Influence of crop edges on movement of generalist predators: a diffusion approach. *Agricultural and Forest Entomology* 4, 21–30.

Botes, A., McGeoch, M.A. & van Rensberg, B.J. (2006) Elephant- and human-induced changes to dung beetle (Coleoptera: Scarabaeidae) assemblages in the Maputaland Centre of Endemism. *Biological Conservation* 130, 573–583.

Bouchard, A.-M., McNeil, J.N. & Brodeur, J. (2007) Invasion of American native lily populations by an alien beetle. *Biological Invasions* 10, 1365–1372.

Bouchard, P., Grebennikov, V.V., Smith, A.B.T. & Douglas, H. (2009) Biodiversity of Coleoptera. In: *Insect Biodiversity: Science and Society* (eds R. Foottit & P. Adler), pp. 265–301. Blackwell Publishing, Oxford.

Boving, A.G. & Craighead, F.C. (1931) *An Illustrated Synopsis of the Principal Larval Forms of the Order Coleoptera*. Brooklyn Entomological Society, Brooklyn, New York.

Boyce, D. & Walters, J. (2001) The conservation of the blue ground beetle in south-west England. *British Wildlife* 13, 101–108.

Brandle, M., Durka, W. & Altmoos, M. (2000) Diversity of surface dwelling beetle assemblages in open-cast lignite mines in Central Germany. *Biodiversity and Conservation* 9, 1297–1311.

Brandmayr, P., Lovei, G.L., Casle, A. & Tagnlianti, A.V. (eds) (2000) *Natural History and Applied Ecology of Carabid Beetles*. Pensoft, Sofia.

Brockerhoff, E.G., Berndt, L.A. & Jactel, H. (2005) Role of exotic pine forest in the conservation of the critically endangered New Zealand ground beetle *Holcaspis brevicula* (Coleoptera: Carabidae). *New Zealand Journal of Ecology* **29**, 37–43.

Brockerhoff, E.G., Bain, J., Kimberley, M. & Knizek, M. (2006) Interception frequency of exotic bark and ambrosia beetles (Coleoptera: Scolytinae) and relationship with establishment in New Zealand and worldwide. *Canadian Journal of Forest Research* **36**, 289–298.

Burbidge, A.A. (1996) Essentials of a good recovery plan. In: *Back from the Brink: Refining the Threatened Species Recovery Process* (eds S. Stephens & S. Maxwell), pp. 55–62. Surrey Beatty & Sons, Chipping Norton.

Burel, F. (1989) Landscape structure effects on carabid beetles spatial patterns in western France. *Landscape Ecology* **2**, 215–226.

Burel, F. & Baudry, J. (1990) Hedgerow networks as habitats for forest species: implications for colonizing abandoned agricultural land. In: *Species Dispersal in Agricultural Habitats* (eds R.G.H. Bunce & D.C. Howard), pp. 238–255. Belhaven Press, London.

Burel, F., Baudry, J., Delette, Y., Petit, S. & Morvan, N. (2000) Relating insect movements to farming systems in dynamic landscapes. In: *Interchange of Insects between Agricultural and Surrounding Landscapes* (eds B. Ekbom, M.E. Irwin & Y. Robert), pp. 5–32. Kluwer Academic Publishers, Dordrecht.

Burke, D. & Goulet, H. (1998) Landscape and area effects on beetle assemblages in Ontario. *Ecography* **21**, 472–479.

Buse, A. (1988) Habitat selection and grouping of beetles (Coleoptera). *Holarctic Ecology* **11**, 241–247.

Buse, A. & Good, J.E.G. (1993) The effects of conifer forest design and management on abundance and diversity of rove beetles (Coleoptera: Staphylinidae): implications for conservation. *Biological Conservation* **64**, 67–76.

Buse, J., Schroder, B. & Assmann, T. (2007) Modelling habitat and spatial distribution of an endangered longhorn beetle: a case study for saproxylic insect conservation. *Biological Conservation* **137**, 372–381.

Buse, J., Ranius, T. & Assmann, T. (2008) An endangered longhorn beetle associated with old oaks and its possible role as an ecosystem engineer. *Conservation Biology* **22**, 329–337.

Bussler, H. & Müller, J. (2009) Vacuum cleaning for conservationists: a new method for inventory of *Osmoderma eremita* (Scop., 1763) (Coleoptera: Scarabaeidae) and other inhabitants of hollow trees in Natura 2000 areas. *Journal of Insect Conservation* **13**, 355–399.

Butterfield, J. (1996) Carabid life-cycle strategies and climate change: a study on an altitude transect. *Ecological Entomology* **21**, 9–16.

Butterfield, J. (1997) Carabid community succession during the forestry cycle in conifer plantations. *Ecography* **20**, 614–625.

Butterfield, J., Luff, M.L., Barnes, M. & Eyre, M.D. (1995) Carabid beetle communities as indicators of conservation potential in upland forests. *Forest Ecology and Management* **79**, 63–77.

Butterflies Under Threat Team (1986) *The Management of Chalk Grassland for Butterflies*. Focus on Nature Conservation no. 17. Nature Conservancy Council, Peterborough.

Caltagirone, L.E. & Doutt, R.L. (1989) The history of the vedalia beetle importation to California and its impact on the development of biological control. *Annual Review of Entomology* **34**, 1–16.

Carpaneto, G.M., Mazziotta, A. & Piattella, E. (2005) Changes in food resources and conservation of scarab beetles: from sheep to dog dung in a green urban area of Rome (Coleoptera, Scarabaeoidea). *Biological Conservation* **123**, 547–556.

Carpaneto, G.M., Mazziotta, A. & Valerio, L. (2007) Inferring species decline from collection records: roller dung beetles in Italy (Coleoptera: Scarabaeidae). *Diversity and Distributions* **13**, 903–919.

Cartagena, M.C. & Galante, E. (2002) Loss of Iberian island tenebrionid beetles and conservation management recommendations. *Journal of Insect Conservation* **6**, 673–681.

Carter, G.A., Seal, M.R. & Haley, T. (1998) Airborne detection of southern pine beetle damage using key spectral bands. *Canadian Journal of Forest Research* **28**, 1040–1045.

Castillo, M.L. & Lobo, J.M. (2004) A comparison of Passalidae (Coleoptera, Lamellicornia) diversity and community structure between primary and secondary tropical forest in Lox Tuxtlas, Veracruz, Mexico. *Biodiversity and Conservation* **13**, 1257–1269.

Caton, B.P., Dobbs, T.T. & Brodel, C.F. (2006) Arrivals of hitchhiking insect pests on international cargo aircraft at Miami International Airport. *Biological Invasions* **8**, 765–785.

Chapman, J.W., Birkett, M.A., Pickett, J.A. & Woodcock, C.M. (2002) Chemical ecology and conservation of the stag beetle *Lucanus cervus*. *Comparative Biochemistry and Physiology A* **132**, S63.

Charrier, S., Petit, S. & Burel, F. (1997) Movements of *Abax parallelepidus* (Coleoptera, Carabidae) in woody habitats of a hedgerow network landscape: a radio-tracing study. *Agriculture, Ecosystems and Environment* **61**, 133–144.

Cheesman, O.D. & Key, R.S. (2007) The extinction of experience: a threat to insect conservation? In: *Insect Conservation Biology* (eds A.J.A. Stewart, T.R. New & O.T. Lewis), pp. 322–348. CABI Publishing, Wallingford.

Chown, S.L. (1992) A preliminary analysis of sub-Antarctic weevil assemblages: local and regional patterns. *Journal of Biogeography* **19**, 87–98.

Chown, S.L. (1994) Historical ecology of sub-Antarctic weevils (Coleoptera: Curculionidae): patterns and processes on isolated islands. *Journal of Natural History* **28**, 411–433.

Chown, S.L., Scholtz, C.H., Klok, C.J., Joubert, F.J. & Coles, K.S. (1995) Ecophysiology, range contraction and survival of a geographically restricted African dung beetle (Coleoptera: Scarabaeidae). *Functional Ecology* **9**, 30–39.

Chung, A.Y.C., Eggleton, P., Speight, M.R., Hammond, P.M. & Chey, V.K. (2000) The diversity of beetle assemblages in different habitat types in Sabah, Malaysia. *Bulletin of Entomological Research* **90**, 475–496.

Clarke, G.M. & Spier, F. (2003) *A Review of the Conservation Status of Selected Australian Non-marine Invertebrates*. Natural Heritage Trust/Environment Australia, Canberra.

Clough, Y., Kruess, A. & Tscharntke, T. (2007) Organic versus conventional arable farming systems: functional grouping helps understand staphylinid response. *Agriculture, Ecosystems and Environment* **118**, 285–290.

Cole, L.J., McCracken, D.I., Dennis, P. *et al.* (2002) Relationships between agricultural management and ecological groups of ground beetles (Coleoptera: Carabidae) on Scottish farmland. *Agriculture, Ecosystems and Environment* **93**, 323–336.

Collinge, S.K., Holyoak, M., Barr, C.B. & Marty, J.T. (2001) Riparian habitat fragmentation and population persistence of the threatened valley elderberry longhorn beetle in central California. *Biological Conservation* **100**, 103–113.

Collins, K.L., Boatman, N.D., Wilcox, A.W. & Chaney, K. (2002) Influence of beetle banks on cereal aphid predation in winter wheat. *Agriculture, Ecosystems and Environment* **93**, 337–350.

Collins, N.M. (1987) *Legislation to Conserve Insects in Europe*. Amateur Entomologist's Society, London.

Connell, J.H. (1978) Diversity in tropical rain forest and coral reefs. *Science* **199**, 1302–1310.

Contreras-Diaz, H.G., Moya, O., Oromi, P. & Juan, C. (2003) Phylogeography of the endangered darkling beetle species of *Pimelia* endemic to Gran Canaria (Canary Islands). *Molecular Ecology* **12**, 2131–2143.

Coope, G.R. (1995) Insect faunas in ice age environments: why so little extinction? In: *Extinction Rates* (eds J.H. Lawton & R.M. May), pp. 55–74. Oxford University Press, Oxford.

Cooper, A., McCann, T., Davidson, R. & Foster, G.N. (2005) Vegetation, water beetles and habitat isolation in abandoned lowland bog drains and peat pits. *Aquatic Conservation: Marine and Freshwater Ecosystems* **15**, 175–188.

Cooper, S.J.B., Hinze, S., Leys, R., Watts, C.H.S. & Humphreys, W.F. (2002) Islands under the desert: molecular systematics and evolutionary origins of stygobitic water beetles (Coleoptera: Dytiscidae) from central Western Australia. *Invertebrate Systematics* **16**, 589–598.

Cooter, J. & Barclay, M.V.L. (eds) (2006) *A Coleopterist's Handbook*, 4th edn. Amateur Entomologist's Society, London.

Coulson, R.N. & Wunneburger, W.F. (2000) Impacts of insects in human-dominated and natural forest landscapes. In: *Invertebrates as Webmasters in Ecosystems* (eds D.C. Coleman & P.F. Hendrix), pp. 271–291. CABI Publishing, Wallingford.

Crowson, R.A. (1981) *The Biology of the Coleoptera*. Academic Press, London.

Culver, D.C. (1982) *Cave Life. Evolution and Ecology*. Harvard University Press, Cambridge, MA.

Dadour, I.R., Cook, D.F. & Neesam, C. (1999) Dispersal of dung containing ivermectin in the field by *Onthophagus taurus* (Coleoptera: Scarabaeidae). *Bulletin of Entomological Research* **89**, 119–123.

Danks, H.V., Wiggins, G.B. & Rosenberg, D.M. (1987) Ecological collections and long-term monitoring. *Bulletin of the Entomological Society of Canada* **19**, 16–18.

Davies, K.F. & Margules, C.R. (1998) Effects of habitat fragmentation on carabid beetles: experimental evidence. *Journal of Animal Ecology* **67**, 460–471.

Davies, Z.G., Tyler, C., Stewart, G.B. & Pullin, A.S. (2008) Are current management recommendations for saproxylic invertebrates effective? A systematic review. *Biodiversity and Conservation* 17, 209–234.

Davis, A.J., Huijbregts, H. & Krikken, J. (2000) The role of local and regional processes in shaping dung beetle communities in tropical forest plantations in Borneo. *Global Ecology and Biogeography* 9, 281–292.

Davis, A.J., Holloway, J.D., Huijbregts, H., Krikken, J., Kirk-Spriggs, A.H. & Sutton, S.L. (2001) Dung beetles as indicators of change in the forests of northern Borneo. *Journal of Applied Ecology* 38, 593–616.

Davis, A.L.V. (2002) Dung beetle diversity in South Africa: influential factors, conservation status, data inadequacies and sampling design. *African Entomology* 10, 53–65.

Davis, A.L.V., Scholtz, C.H. & Chown, S.L. (1999) Species turnover, community boundaries and biogeographical composition of dung beetle assemblages across an altitudinal gradient in South Africa. *Journal of Biogeography* 26, 1039–1055.

Davis, A.L.V., Van Aarde, R.J., Scholtz, C.H. & Delport, J.H. (2002) Increasing representation of localized dung beetles across a chronosequence of regeneration vegetation and natural dune forest in South Africa. *Global Ecology and Biogeography* 11, 191–209.

Day, W.H., Prokrym, D.R., Ellis, D.R. & Chianese, R.J. (1994) The known distribution of the predator *Propylea quatuordecimpunctata* (Coleoptera: Coccinellidae) in the United States, and thoughts on the origin of this species and five other exotic lady beetles in eastern North America. *Entomological News* 105, 24–256.

de Groot, P. & Nott, R. (2001) Evaluation of traps of six different designs to capture pine sawyer beetles (Coleoptera: Cerambycidae). *Agricultural and Forest Entomology* 3, 107–111.

den Boer, P.J. (1971) On the dispersal power of carabid beetles and its possible significance. *Miscellaneous Papers, Landb. Hogeschool, Wageningen* 8, 119–137.

den Boer, P.J. (1987) On the turnover of carabid populations in changing environments. *Acta Phytopatologica et Entomologica Hungarici* 22, 71–83.

den Boer, P.J., Luff, M.L., Mossakowski, D. & Weber, F. (eds) (1986) *Carabid Beetles. Their Adaptations and Dynamics.* Gustav Fisher, Stuttgart and New York.

Dennis, R.L.H., Shreeve, T.G. & Van Dyck, H. (2006) Habitats and resources: the need for a resource-based definition to conserve butterflies. *Biodiversity and Conservation* 15, 1943–1968.

Dennis, R.L.H., Shreeve, T.G. & Sheppard, D.A. (2007) Species conservation and landscape management: a habitat perspective. In: *Insect Conservation Biology* (eds A.J.A. Stewart, T.R. New & O.T. Lewis), pp. 92–126. CABI Publishing, Wallingford.

Desender, K. & Bosmans, R. (1998) Ground beetles (Coleoptera, Carabidae) on set-aside fields in the Campine region and their importance for natural conservation in Flanders (Belgium). *Biodiversity and Conservation* 7, 1485–1493.

Desender, K. & Turin, H. (1989) Loss of habitats and changes in the composition of the ground and tiger beetle fauna in four west European countries since 1950 (Coleoptera: Carabidae, Cicindelidae). *Biological Conservation* 48, 277–294.

De Vries, H.H. & den Boer, P.J. (1990) Survival of populations of *Agonum ericeti* Panz. (Col., Carabidae) in relation to fragmentation of habitats. *Netherlands Journal of Zoology* 40, 484–489.

Didham, R.K., Hammond, P.M., Lawton, J.H., Eggleton, P. & Stork, N.E. (1998a) Beetle species richness responses to tropical forest fragmentation. *Ecological Monographs* 68, 295–323.

Didham, R.K., Lawton, J.H., Hammond, P.M. & Eggleton, P. (1998b) Trophic structure stability and extinction dynamics of beetles (Coleoptera) in tropical forest fragments. *Philosophical Transactions of the Royal Society of London B* 353, 437–451.

Diogo, A.C., Vogler, A.P., Gimenez, A., Gallenga, D. & Galiano, J. (1999) Conservation genetics of *Cicindela deserticoloides*, an endangered tiger beetle endemic to southeastern Spain. *Journal of Insect Conservation* 3, 117–123.

Dixon, A.F.G. (2000) *Insect Predator–Prey Dynamics: Ladybird Beetles and Biological Control.* Cambridge University Press, Cambridge.

Donlan, E.M., Townsend, J.H. & Golden, E.A. (2004) Predation of *Caretta caretta* (Testudines: Cheloniidae) eggs by larvae of *Lanelater sallei* (Coleoptera: Elateridae) on Key Biscayne, Florida. *Caribbean Journal of Science* 40, 415–420.

Doube, B.M. (1990) A functional classification for analysis of the structure of dung beetle assemblages. *Ecological Entomology* 15, 371–383.

Doube, B.M., Macqueen, A., Ridsdill-Smith, T.J. & Weir, T.A. (1991) Native and introduced dung beetles in Australia. In: *Dung Beetle Ecology* (eds I. Hanski & Y. Cambefort), pp. 255–278. Princeton University Press, Princeton, NJ.

Drees, C., Matern., A., Rasplus, J.-Y., Terlutter, H., Assmann, T. & Weber, F. (2008) Microsatellites and allozymes as the genetic memory of habitat fragmentation in populations of the ground beetle *Carabus auronitens* (Col., Carabidae). *Journal of Biogeography* 35, 1937–1949.

du Bus de Warnaffe, G. & Lebrun, P. (2004) Effects of forest management on carabid beetles in Belgium: implications for biodiversity conservation. *Biological Conservation* 118, 219–234.

Duelli, P. (1988) Aphidophaga and the concepts of island biogeography in agricultural areas. In: *Ecology and Effectiveness of Aphidophaga* (eds E. Niemczyk & A.F.G. Dixon), pp. 89–93. SPB Publishing, The Hague.

Duelli, P., Studer, M., Machard, I. & Jakob, S. (1990) Population movements of arthropods between natural and cultivated areas. *Biological Conservation* 54, 193–207.

Dufrene, M. & Legendre, P. (1997) Species assemblages and indicator species: the need for a flexible asymmetrical approach. *Ecological Monographs* 67, 345–366.

Dunn, R.R. (2005) Modern insect extinctions, the neglected majority. *Conservation Biology* 19, 1030–1036.

Dunn, R.R. & Danoff-Burg, J. (2007) Road size and carrion beetle assemblages in a New York forest. *Journal of Insect Conservation* 11, 325–332.

Elkins, R. (2006) Fans exterminate 'Hitler' beetle. *Independent*, 20 August 2006. Available at www.independent.co.uk/news/europe/fans-exterminate-hitler-beetle-412632.html (accessed 5 March 2008).

Emerson, B.C. & Wallis, G.P. (1994) Species status and population genetic structure of the flightless chafer beetles *Prodontria modesta* and *P. bicolorata* (Coleoptera; Scarabaeidae) from South Island, New Zealand. *Molecular Ecology* 3, 339–345.

Erwin, T.L. (1982) Tropical forests: their richness in Coleoptera and other arthropod species. *Coleopterists Bulletin* 36, 74–75.

Erwin, T.L. (1988) The tropical forest canopy: the heart of biotic diversity. In: *Biodiversity* (ed. E.O. Wilson), pp. 123–129. National Academy Press, Washington, DC.

Erwin, T.L. (1997) Biodiversity at its utmost: tropical forest beetles. In: *Biodiversity II. Understanding and Protecting Our Biological Resources* (eds M.L. Reaka-Kudla, D.E. Wilson & E.O. Wilson), pp. 27–40. Joseph Henry Press, Washington, DC.

Erwin, T.L. & Aschero, V. (2004) *Cicindis horni* Bruch (Coleoptera: Carabidae, Cicindini): the fairy shrimp hunting beetle, its way of life on the Salinas Grandes of Argentina. *Zootaxa* 553, 1–16.

Erwin, T.L. & Geraci, C.J. (2009) Amazonian rainforests and their richness of Coleoptera, a dominant life form in the critical zone of the Neotropics. In: *Insect Biodiversity: Science and Society* (eds R. Foottit & P. Adler), pp. 49–67. Blackwell Publishing, Oxford.

Erwin, T.L., Ball, G.E. & Whitehead, D.R. (eds) (1979) *Carabid Beetles: Their Evolution, Natural History, and Classification*. W. Junk, The Hague.

Escobar, F., Halffter, G. & Arellano, L. (2007) From forest to pasture: an evaluation of the influence of environment and biogeography on the structure of dung beetle (Scarabaeinae) assemblages along three altitudinal gradients in the Neotropical region. *Ecography* 30, 193–208.

Evans, A.V. & Bellamy, C.L. (1996) *An Inordinate Fondness for Beetles*. Henry Holt & Co., New York.

Evans, W.G. (1970) *Thalassotrechus barbarae* (Horn) and the Santa Barbara oil spill (Coleoptera: Carabidae). *Pan-Pacific Entomologist* 436, 233–237.

Eversham, B.C. & Telfer, M.C. (1994) Conservation value of roadside verges for stenotopic heathland Carabidae: corridors or refugia? *Biodiversity and Conservation* 3, 538–545.

Eversham, B.C., Roy, D.B. & Telfer, M.G. (1996) Urban, industrial and other manmade sites as analogues of natural habitats for Carabidae. *Annales Zoologicae Fennici* 33, 149–156.

Ewers, R.M. & Didham, R.K. (2008) Pervasive impact of large-scale edge effects on a beetle community. *Proceedings of the National Academy of Sciences USA* 105, 5426–5429.

Eyre, M.D. (2006) A strategic interpretation of beetle (Coleoptera) assemblages, biotopes, habitats and distribution, and the conservation implications. *Journal of Insect Conservation* 10, 151–160.

Eyre, M.D. & Luff, M.L. (2002) The use of ground beetles (Coleoptera: Carabidae) in conservation assessment of exposed riverine sediment habitats in Scotland and northern England. *Journal of Insect Conservation* 6, 25–38.

Eyre, M.D., Lott, D.A. & Garside, A. (1996) Assessing the potential for environmental monitoring using ground beetles (Coleoptera: Carabidae) with riverside and Scottish data. *Annales Zoologicae Fennici* **33**, 157–163.

Eyre, M.D., Luff, M.L. & Phillips, D.A. (2001) The ground beetles (Coleoptera: Carabidae) of exposed riverine sediments in Scotland and northern England. *Biodiversity and Conservation* **10**, 403–426.

Eyre, M.D., Luff, M.L., Staley, J.R. & Telfer, M.G. (2003a) The relationship between British ground beetles (Coleoptera, Carabidae) and land cover. *Journal of Biogeography* **30**, 719–730.

Eyre, M.D., Luff, M.L. & Woodward, J.C. (2003b) Beetles (Coleoptera) on brownfield sites in England: an important conservation resource? *Journal of Insect Conservation* **7**, 223–231.

Eyre, M.D., Foster, G.N., Luff, M.L. & Staley, J.R. (2003c) An investigation into the relationship between water beetle (Coleoptera) distribution and land cover in Scotland and northeast England. *Journal of Biogeography* **30**, 1835–1849.

Eyre, M.D., Foster, G.N., Luff, M.L. & Rushton, S.P. (2006) The definition of British water beetle species pools (Coleoptera) and their relationship to altitude, temperature, precipitation and land cover variables. *Hydrobiologia* **560**, 121–131.

Fairchild, G.W., Faulds, A.M. & Matta, J.F. (2000) Beetle assemblages in ponds: effects of habitat and site age. *Freshwater Biology* **44**, 523–534.

Faithfull, I. (1992) Records of native dung beetles *Onthophagus pexatus* Harold, *O. auritus* Erichson and *O. granulatus* Boheman (Coleoptera: Scarabaeidae) at dog scats and their potential for the biocontrol of dog dung. *Victorian Entomologist* **22**, 105–108.

Farrell, B.D. (1998) 'Inordinate fondness' explained: why are there so many beetles? *Science* **281**, 555–559.

Fayt, P., Machmer, M.M. & Steeger, C. (2005) Regulation of spruce bark beetles by woodpeckers: a literature review. *Forest Ecology and Management* **206**, 1–14.

Fayt, P., Dufrene, M., Branquant, E. *et al.* (2006) Contrasting responses of saproxylic insects to focal habitat resources: the example of longhorn beetles and hoverflies in Belgian deciduous forest. *Journal of Insect Conservation* **10**, 129–150.

Feer, F. & Pincebourde, S. (2005) Diel flight activity and ecological segregation within an assemblage of tropical forest dung and carrion beetles. *Journal of Tropical Ecology* **21**, 21–30.

Ferguson, A.J. & Pearce-Kelly, P. (2004) *Husbandry Guidelines for the Frégate Beetle* Polposipus herculeanus. Federation of Zoological Gardens of Great Britain and Ireland, London.

Finn, J.A. & Gittings, T. (2003) A review of competition in north temperate dung beetle communities. *Ecological Entomology* **28**, 1–13.

Floate, K.D. (1998) Off-target effects of ivermectin on insects and on dung degradation in southern Alberta, Canada. *Bulletin of Entomological Research* **88**, 25–35.

Floren, A. & Linsenmaier, K.E. (2003) How do beetle assemblages respond to anthropogenic disturbances? In: *Arthropods of Tropical Forests: Spatiotemporal Dynamics and Resource Use in the Canopy* (eds Y. Basset, R. Kitching, S. Miller & V. Novotny), pp. 190–197. Cambridge University Press, Cambridge.

Foster, G.N. (1987) The use of Coleoptera records in assessing the conservation status of wetlands. In: *The Use of Invertebrate Community Data in Environmental Assessment* (ed. M.L. Luff), pp. 8–18. Proceedings of a Meeting of the Agricultural Environment Research Group, University of Newcastle upon Tyne.

Foster, G.N. (2000) The aquatic Coleoptera of British saltmarshes: extremes of generalization and specialism. In: *British Saltmarshes* (eds B.R. Sherwood, B.G. Gardiner & T. Hughes), pp. 223–233. Forrest Text, Cardigan.

Foster, G.N. & Eyre, M.D. (1992) *Classification and Ranking of Water Beetle Communities*. Joint Nature Conservation Committee, Peterborough.

Foster, G.N., Foster, A.P., Eyre, M.D. & Bilton, D.T. (1990) Classification of water beetle assemblages in arable fenland and ranking of sites in relation to conservation value. *Freshwater Biology* **22**, 343–354.

Foster, G.N., Nelson, B.H., Bilton, D.T. *et al.* (2006) A classification and evaluation of Irish water beetle assemblages. *Aquatic Conservation: Marine and Freshwater Ecosystems* **2**, 185–208.

Fournier, E. & Loreau, M. (1999) Effects of newly planted hedges on ground-beetle diversity (Coleoptera, Carabidae) in an agricultural landscape. *Ecography* **22**, 87–97.

Fowles, A.P., Alexander, K.N.A. & Key, R.S. (1999) The saproxylic quality index evaluating wooded habitats for the conservation of dead wood in forests. *Coleopterist* **8**, 121–140.

Frampton, G.K., Cilgi, T., Fry, G.L.A. & Wratten, S.D. (1995) Effects of grassy banks on the dispersal of some carabid beetles (Coleoptera, Carabidae) on farmland. *Biological Conservation* 71, 347–355.

Franc, N. (2007) Standing or downed trees: does it matter for saproxylic beetles in temperate oak-rich forest? *Canadian Journal of Forest Research* 37, 2494–2507.

Frankham, R., Ballou, J.D. & Briscoe, D.A. (2002) *Introduction to Conservation Genetics*. Cambridge University Press, Cambridge.

Freitag, R. (1979) Carabid beetles and pollution. In: *Carabid Beetles: Their Evolution, Natural History, and Classification* (eds T.L. Erwin, G.E. Ball & D.R. Whitehead), pp. 507–521. W. Junk, The Hague.

Frith, M. (2000) *Stag Beetle. An Advice Note for its Conservation in London*. London Wildlife Trust, London.

Fuellhaas, U. (2000) Restoration of degraded fen grassland: effects of long-term inundation and water logging on ground beetle populations (Coleoptera, Carabidae). In: *Natural History and Applied Ecology of Carabid Beetles* (eds P. Brandmayr, G.L. Lovei, A. Casle & A.V. Tagnlianti), pp. 251–263. Pensoft, Sofia.

Galante, E., Garcia-Roman, M., Barrera, I. & Galindo, P. (1991) Comparison of spatial distribution patterns of dung-feeding scarabs (Coleoptera: Scarabaeidae, Geotrupidae) in wooded and open pastureland in the Mediterranean 'dehesa' area of the Iberian Peninsula. *Environmental Entomology* 20, 90–97.

Gaston, K.J. (1991) The magnitude of global insect species richness. *Conservation Biology* 5, 283–296.

Gaston, K.J. & McArdle, B.H. (1994) All else is not equal: temporal population variability and insect conservation. In: *Perspectives on Insect Conservation* (eds K.J. Gaston, T.R. New & M.J. Samways), pp. 171–184. Intercept, Andover.

Geertsema, H. (2004) Some notes on *Colophon* (Coleoptera: Lucanidae) and peripherals. *Colophon (Newsletter of the IUCN/SSC Southern African Invertebrates Specialist Group)* no. 4, 1–4.

Geertsema, H. & Owen, C.R. (2007) Notes on the habits and adult behaviour of three red-listed *Colophon* spp. (Coleoptera; Lucanidae) of the Cape Floristic Region, South Africa. *Journal of Insect Conservation* 11, 43–46.

Gibb, H., Hjalten, J., Ball, J.P. et al. (2006) Effects of landscape composition and substrate availability on saproxylic beetles in boreal forests: a study using experimental logs for monitoring assemblages. *Ecography* 29, 191–204.

Gibbs, J.P. & Stanton, E.J. (2001) Habitat fragmentation and arthropod community change: carrion beetles, phoretic mites, and flies. *Ecological Applications* 11, 79–85.

Goka, K. & Kojima, H. (2004) Genetic disturbance caused by commercialisation of stag beetles in Japan. Abstract of paper given at XXII International Congress of Entomology, Brisbane, Queensland, August 2004.

Goka, K., Kojima, H. & Okabe, K. (2004) Biological invasion caused by commercialisation of stag beetles in Japan. *Global Environmental Research* 8, 67–74.

Gongalsky, K.B., Midtgaard, F. & Overgaard, H.J. (2006) Effects of prescribed forest burning on carabid beetles (Coleoptera: Carabidae): a case study in southeastern Norway. *Entomologica Fennica* 17, 325–333.

Gormley, L., Furley, P. & Watt, A. (2007) Distribution of ground-dwelling beetles in fragmented tropical habitats. *Journal of Insect Conservation* 11, 131–139.

Government of British Columbia (2005) *British Columbia's Mountain Pine Beetle Action Plan 2005–2010*. Government of British Columbia, Victoria, Canada.

Greenslade, P. (1999) What entomologists think about listing species for protection. In: *The Other 99%. The Conservation and Biodiversity of Invertebrates* (eds W. Ponder & D. Lunney), pp. 345–349. Royal Zoological Society of New South Wales, Mosman.

Greenwood, M.T. & Wood, P.J. (2003) Effects of seasonal variation in salinity on a population of *Enochrus bicolor* Fabricius 1792 (Coleoptera: Hydrophilidae) and implications for other beetles of conservation interest. *Aquatic Conservation: Marine and Freshwater Ecosystems* 13, 21–34.

Gressitt, J.L. & Hornabrook, R.W. (1977) *Handbook of Common New Guinea Beetles*. Handbook no. 2, Wau Ecology Institute, Wau, Papua New Guinea.

Griffiths, G.J.K., Alexander, C.J., Birt, A. et al. (2005) A method for rapidly mass-marking individually coded ground beetles (Coleoptera: Carabidae) in the field. *Ecological Entomology* 30, 391–396.

Grimbacher, P.S., Catterall, C.P., Kanowski, J. & Proctor, H.C. (2007) Responses of ground-active beetle assemblages to different styles of reforestation on cleared rainforest land. *Biodiversity and Conservation* 16, 2167–2184.

Grove, S.J. (2000a) Trunk window-traps: an effective technique for sampling tropical saproxylic insects. *Memoirs of the Queensland Museum* **46**, 149–160.

Grove, S.J. (2000b) *Impacts of forestry management on saproxylic beetles in the Australian lowland tropics and the development of appropriate indicators of sustainable forest management.* Unpublished PhD thesis, James Cook University, Queensland.

Grove, S.J. (2002) Saproxylic insect ecology and the sustainable management of forests. *Annual Review of Ecology and Systematics* **33**, 1–23.

Grove, S.J. & Stork, N.E. (1999) The conservation of saproxylic insects in tropical forests: a research agenda. *Journal of Insect Conservation* **3**, 67–74.

Grove, S.J. & Stork, N.E. (2000) An inordinate fondness for beetles. *Invertebrate Taxonomy* **14**, 733–739.

Grove, S. & Yaxley, B. (2005) Wildlife habitat strips and native forest ground-active beetle assemblages in plantation nodes in northeast Tasmania. *Australian Journal of Entomology* **44**, 331–343.

Grove, S., Richards, K., Spencer, C. & Yaxley, B. (2006) What lives under large logs in Tasmanian eucalypt forest? *Tasmanian Naturalist* **128**, 86–93.

Gunther, J. & Assmann, T. (2005) Restoration ecology meets carabidology: effects of floodplain restitution on ground beetles (Coleoptera, Carabidae). *Biodiversity and Conservation* **14**, 1583–1606.

Gunther, M.J. & New, T.R. (2003) Exotic pine plantations in Victoria, Australia: a threat to epigaeic beetle (Coleoptera) assemblages? *Journal of Insect Conservation* **7**, 73–84.

Haack, R.A. (2001) Intercepted Scolytidae (Coleoptera) at US ports of entry, 1985–2000. *Integrated Pest Management Reviews* **6**, 253–282.

Haack, R.A. & Cavey, J.F. (2000) Insects intercepted on solid wood packing materials at United States ports-of-entry, 1985–1998. In: *Proceedings of the International Conference on Quarantine Pests for the Forest Sector and their Effects on Foreign Trade.* CORMA (La Corporacion Chilena de la Madera), Concepcion, Chile.

Halffter, G. & Arellano, L. (2002) Response of dung beetle diversity to human-induced changes in a tropical landscape. *Biotropica* **34**, 144–154.

Halffter, G.H. & Matthews, E.G. (1966) The natural history of dung beetles of the subfamily Scarabaeinae (Coleoptera: Scarabaeidae). *Folia Entomologica Mexicana* **12–14**, 1–312.

Halffter, G., Favila, M.E. & Halffter, V. (1992) A comparative study on the structure of the scarab guild in Mexican tropical rain forests and cleared ecosystems. *Folia Entomologica Mexicana* **84**, 134–156.

Halsall, N.B. & Wratten, S.D. (1988) The efficiency of pitfall traps for polyphagous predatory Carabidae. *Ecological Entomology* **13**, 293–299.

Hambler, C. & Speight, M.R. (1996) Biodiversity conservation in Britain: science replacing tradition. *British Wildlife* **6**, 137–147.

Hammond, H.E.J., Langor, D.W. & Spence, J.R. (2001) Early colonization of *Populus* wood by saproxylic beetles (Coleoptera). *Canadian Journal of Forest Research* **31**, 1175–1183.

Hammond, P.M. (1990) Carabids in context. In: *Ground Beetles: Their Role in Ecological and Environmental Studies* (ed. N.E. Stork), pp. 403–409. Intercept, Andover.

Hammond, P.M. (1994) Practical approaches to the estimation of the extent of biodiversity in speciose groups. *Philosophical Transactions of the Royal Society of London* **345**, 119–136.

Hammond, P.M. (2000) Coastal Staphylinidae (rove beetles) in the British Isles, with special reference to saltmarshes. In: *British Saltmarshes* (eds B.R. Sherwood, B.G. Gardiner & T. Hughes), pp. 247–302. Forrest Text. Cardigan.

Hance, T. (2002) Impact of cultivation and crop husbandry practices. In: *The Agroecology of Carabid Beetles* (ed. J.M. Holland), pp. 231–249. Intercept, Andover.

Hanks, L.M. (1999) Influence of the larval host plant on reproductive strategies of cerambycid beetles. *Annual Review of Entomology* **44**, 483–505.

Hanski, I. (1982) Dynamics of regional distribution: the core and satellite species hypothesis. *Oikos* **38**, 210–221.

Hanski, I. (1991a) The dung insect community. In: *Dung Beetle Ecology* (eds I. Hanski & Y. Cambefort), pp. 5–21. Princeton University Press, Princeton, NJ.

Hanski, I. (1991b) North temperate dung beetles. In: *Dung Beetle Ecology* (eds I. Hanski & Y. Cambefort), pp. 75–96. Princeton University Press, Princeton, NJ.

Hanski, I. (1999) *Metapopulation Ecology.* Oxford University Press, Oxford.

Hanski, I. (2005) *The Shrinking World: Ecological Consequences of Habitat Loss.* International Ecology Institute, Oldendorf/Luhe, Germany.

Hanski, I. & Cambefort, Y. (eds) (1991) *Dung Beetle Ecology.* Princeton University Press, Princeton, NJ.

Hanski, I. & Gilpin, M. (1991) Metapopulation dynamics: brief history and conceptual domain. *Biological Journal of the Linnean Society* **42**, 3–16.

Hanski, I. & Koskela, H. (1977) Niche relations amongst dung-inhabiting beetles. *Oecologia* **28**, 203–231.

Hanski, I. & Poyry, J. (2007) Insect populations in fragmented habitats. In: *Insect Conservation Biology* (eds A.J.A. Stewart, T.R. New & O.T. Lewis), pp. 175–202. CABI Publishing, Wallingford.

Harmon, J.P., Stephens, E. & Losey, J. (2007) The decline of native coccinellids (Coleoptera: Coccinellidae) in the United States and Canada. *Journal of Insect Conservation* **11**, 85–94.

Harrison, S. (1991) Local extinctions in a metapopulation context: an empirical evaluation. *Biological Journal of the Linnean Society* **42**, 73–88.

Harrison, S. (1994) Metapopulations and conservation. In: *Large-scale Ecology and Conservation Biology* (eds P.J. Edwards, R.M. May & N.R. Webb), pp. 111–128. Blackwell, Oxford.

Harrison, S. & Taylor, A.D. (1997) Empirical evidence for metapopulation dynamics. In: *Metapopulation Biology* (eds I. Hanski & M.E. Gilpin), pp. 27–42. Academic Press, San Diego.

Hartley, D.J., Koivula, M., Spence, J.R., Pelletier, R. & Ball, G.E. (2007) Effects of urbanisation on ground beetle assemblages (Coleoptera, Carabidae) of grassland habitats in western Canada. *Ecography* **30**, 673–684.

Haslett, J.R. (1997) *Suggested Additions to the Invertebrate Species Listed in Appendix II of the Bern Convention.* Final Report to the Council of Europe, Strasbourg.

Hawkeswood, T., Callister, D.J. & Antram, F. (1991) Collection and export of Australian insects. An analysis of legislative protection and trade in Europe. *TRAFFIC Bulletin* **12**, 41–48.

Hedgren, O. & Weslien, J. (2008) Detecting rare species with random or subjective sampling: a case study of red-listed saproxylic beetles in boreal Sweden. *Conservation Biology* **22**, 212–215.

Hedin, J. & Ranius, T. (2002) Using radio telemetry to study dispersal of the beetle *Osmoderma eremita*, an inhabitant of tree hollows. *Computers and Electronics in Agriculture* **35**, 171–180.

Hedin, J., Isacsson, G., Jonsell, M. & Komonen, A. (2008) Forest fuel piles as ecological traps for saproxylic beetles in oak. *Scandinavian Journal of Forest Research* **23**, 348–357.

Heliola, J., Koivula, M. & Niemela, J. (2001) Distribution of carabid beetles (Coleoptera: Carabidae) across a boreal forest-clearcut ecotone. *Conservation Biology* **15**, 370–377.

Heliovaara, K. & Vaisanen, R. (1993) *Insects and Pollution.* CRC Press, Boca Raton, FL.

Henderson, A., Henderson, D. & Sinclair, J. (2008) *Bugs Alive! A Guide to Keeping Australian Invertebrates.* Museum Victoria, Melbourne.

Hengeveld, R. & Hogeweg, P. (1979) Cluster analysis of the distribution patterns of Dutch carabid species (Col.). In: *Multivariate Methods in Ecological Work* (eds L. Orloci, C.R. Rao & W.M. Stiteler), pp. 65–86. International Cooperative Publishing House, Fairland, MD.

Hespenheide, H.A. (1976) Patterns in the use of single plant hosts by wood-boring beetles. *Oikos* **27**, 161–164.

Hickling, R., Roy, D.B., Hill, J.K. & Thomas, C.D. (2005) A northward shift of range margins in British Odonata. *Global Change Biology* **11**, 502–506.

Hickling, R., Roy, D.B., Hill, J.K., Fox, R. & Thomas, C.D. (2006) The distributions of a wide range of taxonomic groups are expanding polewards. *Global Change Biology* **12**, 450–455.

Hill, C.J. (1996) Habitat specificity and food preferences of an assemblage of tropical dung beetles (Coleoptera: Scarabaeidae: Scarabaeinae). *Journal of Tropical Ecology* **12**, 449–460.

Hochkirch, A., Witzenberger, K.A., Teerling, A. & Niemeyer, F (2007) Translocation of an endangered insect species, the field cricket (*Gryllus campestris* Linnaeus 1758) in northern Germany. *Biodiversity and Conservation* **16**, 3597–3607.

Holland, J.D. (2007) Sensitivity of cerambycid biodiversity indicators to definition of high diversity. *Biodiversity and Conservation* **16**, 2599–2609.

Holland, J.M. (2002) Carabid beetles: their ecology, survival and use in agroecosystems. In: *The Agroecology of Carabid Beetles* (ed. J.M. Holland), pp. 1–40. Intercept, Andover.

Holland, J.M. & Luff, M.L. (2000) The effects of agricultural practices on Carabidae in temperate agroecosystems. *Integrated Pest Management Reviews* **5**, 109–129.

Holland, J.M., Frampton, G.K. & van den Brink, P.J. (2002) Carabids as indicators within temperate arable farming systems: implications from SCARAB and LINK integrated farming systems. In: *The Agroecology of Carabid Beetles* (ed. J.M. Holland), pp. 251–277. Intercept, Andover.

Holldobler, B. & Wilson, E.O. (1990) *The Ants*. Belknap Press of Harvard University Press, Cambridge, MA.

Holliday, N.J. (1992) Carabid fauna (Coleoptera: Carabidae) during postfire regeneration of boreal forest: properties and dynamics of species assemblages. *Canadian Journal of Zoology* **70**, 440–452.

Holloway, B.A. (2007) *Lucanidae (Insecta: Coleoptera)*. Fauna of New Zealand No. 61. Manaaki Whenua Press, Lincoln, Canterbury.

Holmes, P.R., Boyce, D.C. & Reed, D.K. (1993) The ground beetle (Coleoptera: Carabidae) fauna of Welsh peatland biotopes: factors influencing the distribution of ground beetles and conservation implications. *Biological Conservation* **63**, 153–161.

Holyoak, M. & Koch-Munz, M. (2008) The effects of site conditions and mitigation practices on success of establishing the Valley elderberry longhorn beetle and its host plant, blue elderberry. *Environmental Management* **42**, 444–457.

Horgan, F.G. (2005) Aggregated distribution of resources creates competition refuges for rainforest dung beetles. *Ecography* **28**, 603–618.

Howarth, F.G. & Ramsay, G.W. (1991) The conservation of island insects and their habitats. In: *The Conservation of Insects and Their Habitats* (eds N.M. Collins & J.A. Thomas), pp. 71–109. Academic Press, London.

Howden, H.F. (1977) Beetles, beach drift, and island biogeography. *Biotropica* **9**, 53–57.

Howden, H.F. & Howden, A. (2001) Changes through time: a third survey of the Scarabaeinae (Coleoptera: Scarabaeidae) at the Welder Wildlife Refuge. *Coleopterists Bulletin* **55**, 356–362.

Howden, H.F. & Nealis, V.G. (1975) Effects of clearing in a tropical rain forest on the composition of the coprophagous scarab beetle fauna (Coleoptera). *Biotropica* **7**, 77–83.

Huntly, P.M., Van Noort, S. & Hamer, M. (2005) Giving increased value to invertebrates through ecotourism. *South African Journal of Wildlife Research* **35**, 53–62.

Hutton, S.A. & Giller, P.S. (2003) The effects of intensification of agriculture on northern temperate dung beetle communities. *Journal of Applied Ecology* **40**, 994–1007.

Huusela-Veistola, E. (1996) Effects of pesticide use and cultivation techniques on ground beetles (Col., Carabidae) in cereal fields. *Annales Zoologicae Fennici* **33**, 197–205.

Huxel, G.R. (2000) The effect of the Argentine ant on the threatened valley elderberry longhorn beetle. *Biological Invasions* **2**, 81–85.

Hyman, P.S. & Parsons, M.S. (1992) *A Review of the Scarce and Threatened Coleoptera of Great Britain, Part 1*. United Kingdom Nature Conservation Committee, Peterborough.

Hyman, P.S. & Parsons, M.S. (1994) *A Review of the Scarce and Threatened Coleoptera of Great Britain, Part 2*. United Kingdom Nature Conservation Committee, Peterborough.

Iperti, G. (1999) Biodiversity of predaceous Coccinellidae in relation to bioindication and economic importance. *Agriculture, Ecosystems and Environment* **74**, 323–342.

Ishitani, M., Kotze, D.J. & Niemela, J. (2003) Changes in carabid beetle assemblages across an urban-rural gradient in Japan. *Ecography* **26**, 481–489.

IUCN (1994) *IUCN Red List Categories*. IUCN Species Survival Commission, Gland, Switzerland.

IUCN (2001) *IUCN Red List Categories and Criteria: Version 3.1*. IUCN, Gland and Cambridge.

IUCN (2008) *The IUCN Red List of Threatened Species*. Available at http://redlist/org/ (accessed 5 November 2008).

Janzen, D.H. (1997) Wildlife biodiversity management in the tropics. In: *Biodiversity II: Understanding and Protecting Our Biological Resources* (eds M.L. Reaka-Kudla, D.E. Wilson & E.O. Wilson), pp. 411–431. National Academy of Sciences, Washington, DC.

Johansen, S. & Hytteborn, H. (2001) A contribution to the discussion of biota dispersal with drift ice and driftwood in the North Atlantic. *Journal of Biogeography* **28**, 105–115.

Johansson, T., Gibb, H., Hilszczanski, J. et al. (2006) Conservation-oriented manipulations of coarse woody debris affect its value as habitat for spruce-infesting bark and ambrosia beetles (Coleoptera: Scolytinae) in northern Sweden. *Canadian Journal of Forest Research* **36**, 174–185.

Johst, K. & Schops, K. (2003) Persistence and conservation of a consumer-resource metapopulation with local overexploitation of resources. *Biological Conservation* **109**, 57–65.

Jolivet, P. & Hawkeswood, T.J. (1995) *Host Plants of Chrysomelidae of the World. An Essay about the Relationships between the Leaf-beetles and their Food-plants.* Backhuys Publishers, Leiden.

Jonsell, M., Weslien, J. & Ehnstrom, B. (1998) Substrate requirements of red-listed saproxylic invertebrates in Sweden. *Biodiversity and Conservation* 7, 749–764.

Jonsell, M., Schroeder, M. & Larsson, T. (2003) The saproxylic beetle *Bolitophagus reticulatus*: its frequency in managed forests, attraction to volatiles and flight period. *Ecography* 26, 421–428.

Jordal, B.H. & Kirkendall, L.R. (1998) Ecological relationships of a guild of tropical beetles breeding in *Cecropia* petioles in Costa Rica. *Journal of Tropical Ecology* 14, 153–176.

Joyce, K.A., Holland, J.M & Doncaster, C.P. (1999) Influences of hedgerow intersections and gaps on the movement of carabid beetles. *Bulletin of Entomological Research* 89, 523–531.

Kaila, L., Martikainen, P. & Pinttila, P. (1997) Dead trees left in clear-cuts benefit saproxylic Coleoptera adapted to natural disturbances in boreal forest. *Biodiversity and Conservation* 6, 1–18.

Kameoka, S. & Kiyono, H. (2004) *A Survey of the Rhinoceros Beetle and Stag Beetle Market in Japan.* TRAFFIC, Tokyo.

Kanda, N., Yokota, T., Shibata, E. & Sato, H. (2005) Diversity of dung-beetle community in declining Japanese subalpine forest caused by an increasing sika deer population. *Ecological Research* 20, 135–141.

Kane, T.C. & Poulson, T.L. (1976) Foraging by cave beetles: spatial and temporal heterogeneity of prey. *Ecology* 57, 793–800.

Keeney, G.D. & Horn, D.J. (2007) The American burying beetle in Ohio: current status and future directions of the first mainland reintroduction and Ohio's captive rearing program. Available at www.fws.gov/southwest/es/Oklahoma/.../Ohio%20Status%20Captive%Reintroduction%20P (accessed 30 October 2008).

Kinnunen, H., Jarvelainen, K., Pakkala, T. & Tiainen, J. (1996) The effect of isolation on the occurrence of farmland carabids in a fragmented landscape. *Annales Zoologicae Fennici* 33, 165–171.

Kinnunen, H., Tiainen, J. & Tukia, H. (2001) Farmland carabid beetle communities at multiple levels of spatial scale. *Ecography* 24, 189–197.

Kirkendall, L.R. & Odegaard, F. (2007) Ongoing invasions of old-growth tropical forests: establishment of three incestuous beetle species in southern Central America (Curculionidae: Scolytinae). *Zootaxa* 1588, 53–62.

Kirmse, S., Adis, J. & Morawetz, W. (2003) Flowering events and beetle diversity in Venezuela. In: *Arthropods of Tropical Forests: Spatiotemporal Dynamics and Resource Use in the Canopy* (eds Y. Basset, R. Kitching, S. Miller & V. Novotny), pp. 256–268. Cambridge University Press, Cambridge.

Kistner, D.H. (1982) The social insects' bestiary. In: *Social Insects* (ed. H.R. Hermann), vol. 3, pp. 1–244. Academic Press, New York.

Klausnitzer, B. (1983) *Beetles.* Exeter Books, New York.

Klein, B.C. (1989) Effects of forest fragmentation on dung and carrion beetle communities in central Amazonia. *Ecology* 70, 1715–1725.

Klimaszewski, J., Langor, D.W., Work, T.T., Pelletier, G., Hammond, J.H.E. & Germain, C. (2005) The effects of patch harvesting and site preparation on ground beetles (Coleoptera, Carabidae) in yellow birch dominated forests of southeastern Quebec. *Canadian Journal of Forest Research* 35, 2616–2628.

Klimaszewski, J., Langor, D.W., Work, T.T., Hammond, J.H.E. & Savard, K. (2008) Smaller and more numerous harvesting gaps emulate natural forest disturbances: a biodiversity test case using rove beetles (Coleoptera, Staphylinidae). *Diversity and Distributions* 14, 969–982.

Knisley, C.B. & Haines, R.D. (2007) Description and conservation status of a new subspecies of *Cicindela tranquebarica* (Coleoptera: Cicindelidae) from the San Joaquin Valley of California, U.S.A. *Entomological News* 118, 109–126.

Knisley, C.B., Hill, J.M. & Scherer, A.M. (2005) Translocation of threatened tiger beetle *Cicindela dorsalis dorsalis* (Coleoptera: Cicindelidae) to Sandy Hook, New Jersey. *Annals of the Entomological Society of America* 98, 552–557.

Koch, R.L. (2003) The multicolored Asian lady beetle, *Harmonia axyridis*: a review of its biology, uses in biological control, and non-target impacts. *Journal of Insect Science* 3(32), 1–16.

Koch, R.L., Venette, R.C. & Hutchison, W.D. (2006) Predicted impact of an exotic generalist predator on monarch butterfly (Lepidoptera: Nymphalidae) populations: a quantitative risk assessment. *Biological Invasions* 8, 1179–193.

Koch, S.O., Chown, S.L., Davis, A.L.V., Endrody-Younga, S. & van Jaarsveld, A.S. (2000) Conservation strategies for poorly surveyed taxa: a dung beetle (Coleoptera, Scarabaeidae) case study from southern Africa. *Journal of Insect Conservation* 4, 45–56.

Kohler, F. (2000) [*Saproxylic Beetles in Native Forest of the Northern Rhineland. Comparative Studies on the Saproxylic Beetles of Germany and Contribution to German Native Forest Research.*] Naturwaldzellen in Nordrhein-Westfalen VII. Schriftenreihe der Landesanstalz fur Okologie, Bodenordnung und Forsten/Landesampt fur Agraordnung Nordrhein-Westfalen, Recklinghausen.

Koivula, M. (2005) Effects of forest roads on spatial distribution of boreal carabid beetles (Coleoptera: Carabidae). *Coleopterists Bulletin* 59, 465–487.

Koivula, M. & Niemela, J. (2003) Gap felling as a forest harvesting method in boreal forests: responses of carabid beetles (Coleoptera, Carabidae). *Ecography* 26, 179–187.

Koivula, M.J. & Vermuelen, H.J.W. (2005) Highways and forest fragmentation effects on carabid beetles (Coleoptera, Carabidae). *Landscape Ecology* 20, 911–926.

Kolbe, W. (1969) Kafer in Wirkungsbereich der Roten Waldameise. *Entomologische Zeitschrift* 79, 269–278.

Komonen, A. (2007) Are we conserving peripheral populations? An analysis of range structure of longhorn beetles (Coleoptera: Cerambycidae) in Finland. *Journal of Insect Conservation* 11, 281–285.

Kotze, D.J. & O'Hara, R.B. (2003) Species decline: but why? Explanations of carabid beetle (Coleoptera, Carabidae) declines in Europe. *Oecologia* 135, 138–148.

Kotze, D.J. & Samways, M.J. (1999) Invertebrate conservation at the interface between the grassland matrix and natural Afromontane forest fragments. *Biodiversity and Conservation* 8, 1339–1363.

Kozol, A. (1991) *Annual Monitoring of the American Burying Beetle on Block Island*. United States Fish and Wildlife Service, Boston, MA.

Kromp, B. & Steinberger, K.-H. (1992) Grassy field margins and arthropod diversity: a case study on ground beetles and spiders in eastern Austria (Coleoptera: Carabidae; Arachnida: Aranei, Opiliones). *Agriculture, Ecosystems and Environment* 40, 71–93.

Kruger, K. & Scholtz, C.H. (1997) Lethal and sublethal effects of ivermectin on the dung-feeding beetles *Euoniticellus intermedius* (Reiche) and *Onitis alexis* Klug (Coleoptera: Scarabaeidae). *Agriculture, Ecosystems and Environment* 61, 123–131.

Kryger, U., Cole, K.S., Tukker, R. & Scholtz, C.H. (2006) Biology and ecology of *Circellium bacchus* (Fabricius 1781) (Coleoptera Scarabaeidae), a South African dung beetle of conservation concern. *Tropical Zoology* 19, 185–207.

Kuschel, G. & Chown, S.L. (1995) Phylogeny and systematics of the *Ectemnorhinus*-group of genera (Insecta: Coleoptera). *Invertebrate Taxonomy* 9, 841–863.

Lane, S. (2003) Bloody-nosed beetle *Timarcha tenebricosa*. Action for Wildlife: Warwickshire, Coventry and Solihull Local Biodiversity Action Plan. Available at www.warwickshire.gov.uk/biodiversity (accessed September 2008).

Larochelle, A. & Lariviere, M.-C. (2007) *Carabidae (Insecta: Coleoptera): Synopsis of Supraspecific Taxa.* Fauna of New Zealand No. 60. Manaaki Whenua Press, Lincoln, Canterbury.

Larsen, D.J. (1985) Structure in temperate predaceous diving beetle communities (Coleoptera: Dytiscidae). *Holarctic Ecology* 8, 18–32.

Larsen, K.J. & Williams, J.B. (1999) Influence of fire and trapping effort on ground beetles in a reconstructed tallgrass prairie. *Prairie Naturalist* 31, 75–86.

Larsen, T.H. & Forsyth, A. (2005) Trap spacing and transect design for dung beetle biodiversity studies. *Biotropica* 37, 322–325.

Larsen, T.H., Lopera, A. & Forsyth, A. (2008) Understanding trait-dependent community disassembly: dung beetles, density functions, and forest fragmentation. *Conservation Biology* 22, 1288–1298.

Lassau, S.A., Hochuli, D.F., Cassis, G. & Reid, C.A.M. (2005) Effects of habitat complexity on forest beetle diversity: do functional groups respond consistently? *Diversity and Distributions* 11, 73–82.

Lawrence, J.F. & Newton, A.F. (1995) Families and subfamilies of Coleoptera. In: *Biology, Phylogeny and Classification of Coleoptera: Papers Celebrating the 80th Birthday of Roy A. Crowson* (eds J. Pakaluk & S.A. Slipinski), pp. 449–472. Muzeum i Instytut Zoologii PAN, Warsaw.

Lee, J.C. & Landis, D.A. (2002) Non-crop habitat management for carabids. In: *The Agroecology of Carabid Beetles* (ed. J.M. Holland), pp. 279–304. Intercept, Andover.

Lee, M.S.Y. (2000) A worrying systematic decline. *Trends in Ecology and Evolution* 15, 346.

Leggatt, J. (2003) Lord Howe's stolen rare beetles go home. *The Sun-Herald*. Available at http://www.smh.com.au/articles/2003/01/04/1041566268552.html (accessed 15 October 2008).

Lemieux, J.P. & Lindgren, B.S. (2004) Ground beetle responses to patch retention harvesting in high elevation forests of British Columbia. *Ecography* 27, 557–566.

Lewis, O.T. & Basset, Y. (2007) Insect conservation in tropical forests. In: *Insect Conservation Biology* (eds A.J.A. Stewart, T.R. New & O.T. Lewis), pp. 34–56. CABI Publishing, Wallingford.

Leys, R. & Watts, C.H. (2008) Systematics and evolution of the Australian subterranean hydroporine diving beetles (Dytiscidae), with notes on *Carabhydrus*. *Invertebrate Systematics* 22, 217–225.

Lichtwardt, R.W., White, M.M., Cafaro, M.J. & Misra, J.K. (1999) Fungi associated with passalid beetles and their mites. *Mycologia* 91, 694–792.

Liebherr, J.K. & Kruschelnycky, P.D. (2007) Unfortunate encounters? Novel interactions of native *Mecyclothorax*, alien *Trechus obtusus* (Coleoptera: Carabidae), and Argentine ant (*Linepithema humile*, Hymenoptera: Formicidae) across a Hawaiian landscape. *Journal of Insect Conservation* 11, 61–73.

Liebherr, J.K. & Zimmerman, E.C. (2000) Hawaiian Carabidae (Coleoptera). Part 1, Introduction and tribe Platynini. *Insects of Hawaii* 16, 16–494.

Lin, Y.-C., James, R. & Dolman, P.M. (2007) Conservation of heathland beetles (Coleoptera, Carabidae): the value of lowland coniferous plantations. *Biodiversity and Conservation* 16, 1337–1358.

Lindhe, A., Lindelow, A. & Asenblad, N. (2005) Saproxylic beetles in standing dead wood density in relation to substrate sun-exposure and diameter. *Biodiversity and Conservation* 14, 3033–3053.

Linsley, E.G. (1959) Ecology of Cerambycidae. *Annual Review of Entomology* 4, 99–138.

Liu, Y., Axmacher, J.C., Wang, C., Li, L. & Yu, Z. (2007) Ground beetle (Coleoptera: Carabidae) inventories: a comparison of light and pitfall traps. *Bulletin of Entomological Research* 97, 577–583.

Lobo, J.G., Lumaret, J.-P. & Jay-Robert, P. (2002) Modelling the species richness distribution of French dung beetles (Coleoptera, Scarabaeidae) and delimiting the predictive capacity of different groups of explanatory variables. *Global Ecology and Biogeography* 11, 265–277.

Lobo, J.G., Jay-Robert, P. & Lumaret, J.-P. (2004) Modelling the species richness distribution for French Aphodiinae (Coleoptera, Scarabaeoidea). *Ecography* 27, 145–156.

Lobo, J.M., Verdu, J.R. & Numa, C. (2006) Environmental and geographical factors affecting the Iberian distribution of flightless *Jekelius* species (Coleoptera: Geotrupidae). *Diversity and Distributions* 12, 179–188.

Lobo, J.M., Baselga, A., Hortal, J., Jimenez-Valverde, A. & Gomez, J.F. (2007) How does the knowledge about the spatial distribution of Iberian dung beetles accumulate over time? *Diversity and Distributions* 13, 772–780.

Logan, J.A. & Powell, J.A. (2001) Ghost forests, global warming, and the mountain pine beetle (Coleoptera; Scolytidae). *American Entomologist* 47, 160–172.

Lomolino, M.V., Creighton, J.C., Schnell, G.D. & Certain, D.L. (1995) Ecology and conservation of the endangered American burying beetle (*Nicrophorus americanus*). *Conservation Biology* 9, 605–614.

Louda, S.M. (1998) Population growth of *Rhinocyllus conicus* (Coleoptera: Curculionidae) on two species of native thistles in prairie. *Environmental Entomology* 27, 834–841.

Louda, S.M., Arnett, A.E., Rand, T.A. & Russell, F.L. (2003) Invasiveness of some biological control insects and adequacy of their ecological risk assessment and regulation. *Conservation Biology* 17, 73–82.

Lovei, G.L. & Cartellieri, M. (2000) Ground beetles (Coleoptera, Carabidae) in forest fragments of the Manawatu, New Zealand: collapsed assemblages? *Journal of Insect Conservation* 4, 239–244.

Lovei, G.L. & Sunderland, K.D. (1996) Ecology and behavior of ground beetles (Coleoptera; Carabidae). *Annual Review of Entomology* 41, 231–256.

Lovei, G.L., Magura, T., Tothmeresz, B. & Kodobocz, V. (2006) The influence of matrix and edges on species richness patterns of ground beetles (Coleoptera: Carabidae) in habitat islands. *Global Ecology and Biogeography* 15, 283–289.

Lucky, A., Erwin, T.L. & Witman, J.D. (2002) Temporal and spatial diversity and distribution of arboreal Carabidae (Coleoptera) in a western Amazonian rain forest. *Biotropica* 34, 376–386.

Luff, M.L. (1987) Biology of polyphagous ground beetles in agriculture. *Agricultural Zoology Reviews* 2, 237–278.

Luff, M.L. (1993) *The Carabidae (Coleoptera) Larvae of Fennoscandia and Denmark.* Fauna Entomologica Scandinavica 27. E.J. Brill, Leiden.

Luff, M.L. & Eyre, M.D. (2000) Factors affecting the ground beetles (Coleoptera: Carabidae) of some British coastal habitats. In: *British Saltmarshes* (eds B.R. Sherwood, B.G. Gardiner & T. Hughes), pp. 235–245. Forrest Text, Cardigan.

Lumaret, J.P., Galante, E., Lumberas, C. *et al.* (1993) Field effects of ivermectin residues on dung beetles. *Journal of Applied Ecology* **30**, 428–436.

Lundkvist, E., Landin, J. & Millberg, P. (2001) Diving beetle (Dytiscidae) assemblages along environmental gradients in an agricultural landscape in southeastern Sweden. *Wetlands* **21**, 48–58.

Lundkvist, E., Landin, J. & Karlsson, F. (2002) Dispersing diving beetles (Dytiscidae) in agricultural and urban landscapes in south-eastern Sweden. *Annales Zoologicae Fennici* **39**, 109–123.

McFarlane, B.L. & Witson, D.O.T. (2008) Perceptions of ecological risk associated with mountain pine beetle (*Dendroctonus ponderosae*) infestations in Banff and Kootenay National Parks of Canada. *Risk Analysis* **28**, 203–212.

McGeoch, M.A. (1998) The selection, testing and application of terrestrial insects as bioindicators. *Biological Reviews* **73**, 181–201.

McGeoch, M.A., Van Rensberg, B.J. & Botes, A. (2002) The verification and application of bioindicators: a case study of dung beetles in a savanna ecosystem. *Journal of Applied Ecology* **39**, 661–672.

McGeoch, M.A., Schroeder, M., Ekbom, B. & Larson, S. (2007) Saproxylic beetle diversity in a managed boreal forest: importance of stand characteristics and forestry conservation measures. *Diversity and Distributions* **13**, 418–429.

McGuinness, C. (2001) *The Conservation Requirements of New Zealand's Nationally Threatened Invertebrates*. Threatened Species Occasional Publications No. 20. Department of Conservation, Wellington.

McGuinness, C. (2002) *Threatened Carabid Beetles Recovery Plan (2002–2007)*. Department of Conservation, Wellington.

McGuinness, C.A. (2007) Carabid beetle (Coleoptera: Carabidae) conservation in New Zealand. *Journal of Insect Conservation* **11**, 31–41.

McIntire, E.J.B. & Fortin, M.-J. (2006) Structure and function of wildfire and mountain pine beetle forest boundaries. *Ecography* **29**, 309–318.

McLean, I.F.G. & Speight, M.C.D. (1993) Saproxylic invertebrates: the European context. In: *Dead Wood Matters: the Ecology and Conservation of Saproxylic Invertebrates in Britain* (eds K.J. Kirby & C.M. Drake), pp. 21–32. English Nature Science no. 7. English Nature, Peterborough.

Maddison, D.R. (1985) The discovery of *Gehringia olympiae* Darlington (Coleoptera: Carabidae). In: *Taxonomy, Phylogeny and Zoogeography of Beetles and Ants* (ed. G.E. Ball), pp. 35–37. W. Junk, Dordrecht.

Mader, H.J. (1984) Animal habitat isolation by roads and agricultural fields. *Biological Conservation* **29**, 81–96.

Maeto, K., Sato, S. & Miyata, H. (2002) Species diversity of longicorn beetles in humid warm-temperate forests: the impact of forest management practices on old-growth forest species in southwestern Japan. *Biodiversity and Conservation* **11**, 1919–1937.

Magagula, C.N. & Samways, M.J. (2001) Maintenance of ladybeetle diversity across a heterogeneous African agricultural/savanna land mosaic. *Biodiversity and Conservation* **10**, 209–222.

Magura, T. & Tothmeresz, B. (1997) Edge effect on carabids in an oak–hornbeam forest at the Aggtelek National Park (Hungary). *Acta Phytopatologica et Entomologica Hungarici* **33**, 379–387.

Magura, T., Kodobocz, V. & Tothmeresz, B. (2001) Effects of habitat fragmentation on carabids in forest patches. *Journal of Biogeography* **28**, 129–138.

Majerus, M.E.N. (1994) *Ladybirds*. New Naturalist Series no. 81. Collins, London.

Majka, C.G., Nornha, C. & Smith, M. (2006) Adventive and native Byrrhidae (Coleoptera) newly recorded from Prince Edward Island, Canada. *Zootaxa* **1168**, 21–30.

Major, R.E., Smith, D., Cassis, G., Gray, M. & Colgan, D.J. (1999) Are roadside strips important reservoirs of invertebrate diversity? A comparison of the ant and beetle faunas of roadside strips and large remnant woodlands. *Australian Journal of Zoology* **47**, 611–624.

Makino, S., Goto, H., Hasegawa, M. *et al.* (2007) Degradation of longicorn beetle (Coleoptera, Cerambycidae, Disteniidae) fauna caused by conversion from broad-leaved to man-made conifer stands of *Cryptomeria japonica* (Taxodiaceae) in central Japan. *Ecological Research* **22**, 373–381.

Malausa, J.C. & Drescher, J. (1991) The project to rescue the Italian ground beetle *Chrysocarabus olympiae*. *International Zoo Yearbook* **30**, 75–79.

Mankin, R.W., Smith, M.T., Tropp, J.M., Atkinson, E.B. & Jong, D.Y. (2008) Detection of *Anoplophora glabripennis* (Coleoptera: Cerambycidae) larvae in different host trees and tissues by automated analyses of sound-impulse frequency and temporal patterns. *Journal of Economic Entomology* **101**, 838–849.

Marris, J.W.M. (2000) The beetle (Coleoptera) fauna of the Antipodes Islands, with comments on the impact of mice; and an annotated checklist of the insect and arachnid fauna. *Journal of the Royal Society of New Zealand* 30, 169–195.

Martikainen, P. & Kaila, L. (2004) Sampling saproxylic beetles: lessons from a 10-year monitoring study. *Biological Conservation* 120, 171–181.

Martikainen, P. & Kouki, J. (2003) Sampling the rarest: threatened beetles in boreal forest biodiversity inventories. *Biodiversity and Conservation* 12, 1815–1831.

Martikainen, P., Kaila, L. & Haila, Y. (1998) Threatened beetles in white-backed woodpecker habitats. *Conservation Biology* 12, 293–301.

Martikainen, P., Siitonen, J., Punttila, P., Kaila, L. & Mauh, J. (2000) Species richness of Coleoptera in mature managed and old-growth boreal forest in southern Finland. *Biological Conservation* 94, 199–209.

Matern, A., Drees, C., Meyer, H. & Assmann, T. (2008) Population ecology of the rare carabid beetle *Carabus variolosus* (Coleoptera: Carabidae) in north-west Germany. *Journal of Insect Conservation* 12, 591–601.

Masayasu, K. (2005) Flickering lights show a healthy natural environment. *Nipponia* no. 32, 15 March 2005 (in Japanese). English version available at http://web-japan.org/nipponia/n=ipponia32/en/animal/animal01.html (accessed 6 June 2008).

Mawdsley, J.R. (2007) Ecology, distribution, and conservation biology of the tiger beetle *Cicindela patruela consentanea* Dejean (Coleoptera: Carabidae: Cicindelinae). *Proceedings of the Entomological Society of Washington* 109, 17–28.

Mawdsley, J.R. (2008) Use of remote sensing tools to expedite surveys for rare tiger beetles (Insecta: Coleoptera: Cicindelidae). *Journal of Insect Conservation* 12, 689–693.

Mawdsley, J.R. & Sithole, H. (2007) Dry season ecology of riverine tiger beetles in Kruger National Park, South Africa. *African Journal of Ecology* 46, 126–131.

Mawdsley, N.A. & Stork, N.E. (1995) Species extinctions in insects: ecological and biogeographical considerations. In: *Insects in a Changing Environment* (eds R. Harrington & N.E. Stork), pp. 321–369. Academic Press, London.

Meads, M.J. (1995) Translocation of New Zealand's endangered insects as a tool for conservation. In: *Reintroduction Biology of Australian and New Zealand Fauna* (ed. M. Serena), pp. 53–56. Surrey Beatty & Sons, Chipping Norton.

Meggs, J.M. (2002) *Surveys to Determine the Presence/absence of* Lissotes menalcas *(Mt Mangana Stag Beetle) at the Proposed Integrated Timber Processing Site (Southwood), Lonnonvale*. Forestry Tasmania, Hobart.

Meggs, J.M. & Munks, S.A. (2003) Distribution, habitat characteristics and conservation requirements of a forest-dependent threatened invertebrate *Lissotes latidens* (Coleoptera: Lucanidae). *Journal of Insect Conservation* 7, 137–152.

Meggs, J.M. & Taylor, R.J. (1999) Distribution and conservation status of the Mt Mangana stag beetle, *Lissotes menalcas* (Coleoptera: Lucanidae). *Papers and Proceedings of the Royal Society of Tasmania* 1333, 23–28.

Meggs, J.M., Munks, S.A. & Corkrey, R. (2003) The distribution and habitat characteristics of a threatened lucanid beetle *Hoplogonus simsoni* in north-east Tasmania. *Pacific Conservation Biology* 9, 172–185.

Meggs, J.M., Munks, S.A., Corkrey, R. & Richards, K. (2004) Development and evaluation of predictive habitat models to assist the conservation planning of a threatened lucanid beetle, *Hoplogonus simsoni*, in north-east Tasmania. *Biological Conservation* 118, 501–511.

Meijer, J. (1974) A comparative study of the immigration of carabids (Coleoptera: Carabidae) into a new polder. *Oecologia* 16, 185–208.

Michaels, K.F. (2007) Using staphylinid and tenebrionid beetles as indicators of sustainable landscape management in Australia: a review. *Australian Journal of Experimental Agriculture* 47, 435–449.

Michaels, K. & Bornemissza, G. (1999) Effects of clearfell harvesting on lucanid beetles (Coleoptera: Lucanidae) in wet and dry sclerophyll forests in Tasmania. *Journal of Insect Conservation* 3, 85–95.

Michaels, K.F. & McQuillan, P.B. (1995) Impact of commercial forest management on geophilous carabid beetles (Coleoptera: Carabidae) in tall, wet *Eucalyptus obliqua* forest in southern Tasmania. *Australian Journal of Ecology* 20, 316–323.

Miller, S.E. (1997) Late Quaternary insects of Rancho La Brea, California, U.S.A. *Quaternary Proceedings* 5, 185–191.

Moretti, M. & Barbalat, S. (2004) The effects of wildfires on wood-eating beetles in deciduous forests on the southern slopes of the Swiss Alps. *Forest Ecology and Management* 187, 85–103.

Moretti, M., Obrist, M.K. & Duelli, P. (2004) Arthropod biodiversity after forest fires: winners and losers in the winter fire regime of the southern Alps. *Ecography* 27, 173–186.

Moser, J.C., Konrad, H., Kirisits, T. & Carta, L.K. (2005) Phoretic mites and nematode associates of *Scolytus multistriatus* and *Scolytus pygmaeus* (Coleoptera: Scolytidae) in Austria. *Agricultural and Forest Entomology* 7, 169–177.

Muirhead, J.R., Leung, B., van Overdijk, C. *et al.* (2006) Modelling local and long-distance dispersal of invasive emerald ash borer *Agrilus planipennis* (Coleoptera) in North America. *Diversity and Distributions* 12, 71–79.

Müller, J., Bussler, H. & Kneib, T. (2008) Saproxylic beetle assemblages related to silvicultural management intensity and stand structures in a beech forest in southern Germany. *Journal of Insect Conservation* 12, 107–124.

Muller, M. & Job, H. (2009) Managing natural disturbance in protected areas: tourists' attitudes towards the bark beetle in a German national park. *Biological Conservation* 142, 375–383.

Munks, S.A., Richards, K., Meggs, J., Wapstra, M. & Corkrey, R. (2004) Distribution, habitat and conservation of two threatened stag beetles, *Hoplogonus bornemisszai* and *H. vanderschoori* (Coleoptera: Lucanidae) in north-east Tasmania. *Australian Zoologist* 32, 586–596.

Muona, J. (1999) Trapping beetles in boreal coniferous forest: how many species do we miss. *Fennia* 177, 11–16.

New, T.R. (1999) Descriptive taxonomy as a facilitating discipline in invertebrate conservation. In: *The Other 99%. The Conservation and Biodiversity of Invertebrates* (eds W. Ponder & D. Lunney), pp. 154–158. Royal Zoological Society of New South Wales, Mosman.

New, T.R. (2000) The conservation of a discipline: traditional taxonomic skills in insect conservation. *Journal of Insect Conservation* 4, 211–213.

New, T.R. (2005) 'Inordinate fondness': a threat to beetles in south-east Asia? *Journal of Insect Conservation* 9, 147–150.

New, T.R. (2007) Recovery plans for insects: what should they contain, and what should they achieve? *Journal of Insect Conservation* 11, 321–324.

New, T.R. (2008) Conserving narrow range endemic insects in the face of climate change: options for some Australian butterflies. *Journal of Insect Conservation* 12, 585–589.

New, T.R. (2009) *Insect Species Conservation*. Cambridge University Press, Cambridge.

Nichols, E., Larsen, T., Spector, S. *et al.* (2007) Global dung beetle response to tropical forest modification and fragmentation: a quantitative literature review and meta-analysis. *Biological Conservation* 137, 1–19.

Niemela, J. (1993) Mystery of the missing species: species-abundance distribution of boreal ground-beetles. *Annales Zoologicae Fennici* 30, 169–172.

Niemela, J. (1996) From systematics to conservation: carabidologists do it all. *Annales Zoologicae Fennici* 33, 1–4.

Niemela, J. (1997) Invertebrates and boreal forest management. *Conservation Biology* 111, 601–610.

Niemela, J.K. & Spence, J.R. (1994) Distribution of forest dwelling carabids (Coleoptera): spatial scale and the concept of communities. *Ecography* 17, 166–175.

Niemela, J. & Spence, J.R. (1999) Dynamics of local expansion by an introduced species: *Pterostichus melanarius* Ill. (Coleoptera, Carabidae) in Alberta, Canada. *Diversity and Distributions* 5, 121–127.

Niemela, J., Haila, Y., Halme, E., Pajunen, T., Punttila, P. & Tukia, H. (1987) Habitat preferences and conservation status of *Agonum mannerheimii* Dej. in Hame, southern Finland. *Notulae Entomologicae* 67, 175–179.

Niemela, J., Spence, J.R., Langor, D., Haila, Y. & Tukia, H. (1994) Logging and boreal ground-beetle assemblages on two continents: implications for conservation. In: *Perspectives on Insect Conservation* (eds K.J. Gaston, T.R. New & M.J. Samways), pp. 29–50. Intercept, Andover.

Niemela, J., Spence, J.R. & Carcamo, H. (1997) Establishment and interactions of carabid populations: an experiment with native and introduced species. *Ecography* 20, 643–652.

Niemela, J., Kotze, J., Ashworth, A. *et al.* (2000) The search for common anthropogenic impacts on biodiversity: a global network. *Journal of Insect Conservation* 4, 3–9.

Niemela J., Kotze, D.J., Venn, S. *et al.* (2002) Carabid beetle assemblages (Coleoptera: Carabidae) across urban–rural gradients: an international comparison. *Landcape Ecology* 17, 387–401.

Niemela, J., Koivula, M. & Kotze, D.J. (2007) The effects of forestry on carabid beetles (Coleoptera: Carabidae) in boreal forests. *Journal of Insect Conservation* 11, 5–18.

Nilsson, A.N., Elmberg, J. & Sjoberg, K. (1994) Abundance and species richness patterns of predaceous diving beetles (Coleoptera: Dytiscidae) in Swedish lakes. *Journal of Biogeography* 21, 197–206.

Nitterus, K., Astrom, M. & Gunnarsson, B. (2007) Commercial harvest of logging residue in clear-cuts affects the diversity and community composition of ground beetles (Coleoptera: Carabidae). *Scandinavian Journal of Forest Research* 22, 231–240.

Niwa, C.G. & Peck, R.W. (2002) Influence of prescribed fire on carabid beetle (Carabidae) and spider (Araneae) assemblages in forest litter in southwestern Oregon. *Environmental Entomology* 31, 785–796.

Noordijk, J., Prins, D., de Jonge, M. & Vermuelen, R. (2006) Impact of a road on the movements of two ground beetle species (Coleoptera: Carabidae). *Entomologica Fennica* 17, 276–283.

Nurnberger, B. & Harrison, R.G. (1995) Spatial population structure in the whirligig beetle *Dineutes assimilis*: evolutionary inferences based on mitochondrial DNA and field data. *Evolution* 49, 266–275.

Obrycki, J.J. & Kring, T.J. (1998) Predaceous Coccinellidae in biological control. *Annual Review of Entomology* 43, 295–321.

Obrycki, J.J., Elliot, N.C. & Giles, K.L. (2000) Coccinellid introductions: potential for evaluation of nontarget effects. In: *Non-target Effects of Biological Control* (eds P.A. Follet & J.J. Duan), pp. 127–145. Kluwer Academic Publishers, Boston, MA.

Odegaard, F. (2003) Species richness, taxonomic composition and host specificity of phytophagous beetles in the canopy of a tropical dry forest in Panama. In: *Arthropods of Tropical Forests: Spatiotemporal Dynamics and Resource Use in the Canopy* (eds Y. Basset, R. Kitching, S. Miller & V. Novotny), pp. 220–236. Cambridge University Press, Cambridge.

Ohsawa, M. (2004a) Comparison of Elaterid biodiversity among larch plantations, secondary forests, and primary forests in the central mountainous region of Japan. *Annals of the Entomological Society of America* 87, 77–774.

Ohsawa, M. (2004b) Species richness of Cerambycidae in larch plantations and natural broad-leaved forests of the central mountainous region of Japan. *Forest Ecology and Management* 189, 375–385.

Ohsawa, M. (2005) Species richness and composition of Curculionidae (Coleoptera) in a conifer plantation, secondary forest, and old-growth forest in the central mountainous region of Japan. *Ecological Research* 20, 632–645.

Okabe, K. & Goka, K. (2008) Potential impacts on Japanese fauna of canestriniid mites (Acari: Astigmata) accidentally introduced with pet lucanid beetles from Southeast Asia. *Biodiversity and Conservation* 17, 71–81.

Oliff, A.S. (1889) The insect fauna of Lord Howe Island. *Australian Museum Memoirs* 2, 77–98.

Omland, K.S. (2002) Larval habitat and reintroduction site selection for *Cicindela puritana* in Connecticut. *Northeastern Naturalist* 9, 433–450.

Omland, K.S. (2004) Puritan tiger beetle (*Cicindela puritana*) on the Connecticut River. Habitat management and translocation alternatives. In: *Species Conservation and Management: Case Studies* (eds H.R. Akcakaya, M.A. Burgman, O. Kindcall, C.C. Wood, P. Sjogren-Gulve, J.S. Hatfield & M.A. McCarthy), pp. 137–148. Oxford University Press, Oxford.

Oppenheimer, J.R. & Begum, J. (1978) Ecology of some dung beetles (Scarabaeidae and Aphodidae) in Dhaka district. *Bangladesh Journal of Zoology* 6, 23–29.

Orgeas, J. & Andersen, A.N. (2001) Fire and biodiversity: responses of grass-layer beetles to experimental fire regimes in an Australian tropical savanna. *Journal of Applied Ecology* 38, 49–62.

Ottesen, P.S. (1996) Niche segregation of terrestrial alpine beetles (Coleoptera) in relation to environmental gradients and phenology. *Journal of Biogeography* 23, 353–369.

Painter, D. (1999) Macroinvertebrate distributions and the conservation value of aquatic Coleoptera, Mollusca and Odonata in the ditches of traditionally managed and grazing fen at Wicken Fen, UK. *Journal of Applied Ecology* 36, 33–48.

Paivinen, J., Ahlroth, P., Kaitala, V., Kotiaho, J.S., Suhonen, J. & Virola, T. (2003) Species richness and regional distribution of myrmecophilous beetles. *Oecologia* 134, 587–595.

Paivinen, J., Ahlroth, P., Kaitala, V. & Suhonen, J. (2004) Species richness, abundance and distribution of myrmecophilous beetles in nests of *Formica aquilonia* ants. *Annales Zoologicae Fennici* 41, 447–454.

Paulay, G. (1985) Adaptive radiation on an isolated oceanic island: the Cryptorhynchinae (Curculionidae) of Rapa revisited. *Biological Journal of the Linnean Society* 26, 95–187.

Pearce-Kelly, P., Morgan, R., Honan, P. *et al.* (2007) The conservation value of insect breeding programmes: rationale, evaluation tools and example programme case studies. In: *Insect Conservation Biology* (eds A.J.A. Stewart, T.R. New & O.T. Lewis), pp. 57–75. CABI Publishing, Wallingford.

Pearson, D.L. & Cassola, F. (1992) World-wide species richness of tiger beetles (Coleoptera: Cicindelidae): indicator taxon for biodiversity studies. *Conservation Biology* 6, 376–391.

Pearson, D.L. & Cassola, F. (2007) Are we doomed to repeat history? A model of the past using tiger beetles (Coleoptera: Cicindelidae) and conservation biology to anticipate the future. *Journal of Insect Conservation* 11, 47–59.

Pearson, D.L. & Vogler, A.P. (2001) *Tiger Beetles: the Evolution, Ecology and Diversity of the Cicindelids.* Cornell University Press, Ithaca, NY.

Pearson, D.L., Knisley, C.B. & Kazilek, C.J. (2006) *A Field Guide to the Tiger Beetles of the United States and Canada.* Oxford University Press, New York.

Peck, S.B. (1997) *Ammophorus insularis* in Hawaii: a Galapagos Island species immigrant to Hawaii (Coleoptera: Tenebrionidae). *Bishop Museum Occasional Papers* 49, 26–29.

Peck, S.B. (1998) A summary of diversity and distribution of the obligate cave-inhabiting faunas of the United States and Canada. *Journal of Cave and Karst Studies* 60, 18–26.

Peck, S.B. (2006) *The Beetles of the Galapagos Islands, Ecuador: Evolution, Ecology and Diversity (Insecta: Coleoptera).* National Research Council of Canada, Ottawa.

Peck, S.B. & Forsyth, A. (1982) Composition, structure and competitive behaviour in a guild of Ecuadorian rain forest dung beetles (Coleoptera: Scarabaeidae). *Canadian Journal of Zoology* 60, 1624–1634.

Peltonen, M. & Heliovaara, K. (1998) Incidence of *Xylechinus pilosus* and *Cryphalus saltuarius* (Scolytidae) in forest-clearcut edges. *Forest Ecology and Management* 103, 141–147.

Peltonen, M. & Heliovaara, K. (1999) Attack density and breeding success of bark beetles (Coleoptera: Scolytidae) at different distances from forest-clearcut edge. *Agricultural and Forest Entomology* 1, 237–242.

Peltonen, M., Heliovaara, K., Vaisanen, R. & Keronen, J. (1998) Bark beetle diversity at different spatial scales. *Ecography* 21, 510–517.

Peters, R.L. & Darling, J.D.S. (1985) The greenhouse effect and nature reserves. *BioScience* 35, 707–726.

Piel, F., Gilbert, M., De Canniere, C. & Gregoire, J.-C. (2008) Coniferous round wood imports from Russia and Baltic countries to Belgium. A pathway analysis for assessing risks of exotic pest insect introductions. *Diversity and Distributions* 14, 318–328.

Quintero, I. & Roslin, T. (2005) Rapid recovery of dung beetle communities following habitat fragmentation in central Amazonia. *Ecology* 86, 3303–3311.

Rabinowitz, D.S., Cairns, S. & Dillon, T. (1986) Seven forms of rarity and their frequency in the flora of the British Isles. In: *Conservation Biology: the Science of Scarcity and Diversity* (ed. M.E. Soulé), pp. 182–204. Sinauer Associates, Sunderland, MA.

Raffa, K.F. & Berryman, A.A. (1987) Interacting selection pressures in conifer-bark beetle systems: a basis for reciprocal adaptation? *American Naturalist* 129, 234–262.

Rainio, J. & Niemela, J. (2003) Ground beetles (Coleoptera: Carabidae) as bioindicators. *Biodiversity and Conservation* 12, 487–506.

Raithel, C. (1991) *American Burying Beetle (*Nicrophorus americanus*) Recovery Plan.* United States Fish and Wildlife Service, Newton Corner, MA.

Ranius, T. (2000) Minimum viable metapopulation size of a beetle, *Osmoderma eremita*, living in tree hollows. *Animal Conservation* 3, 37–43.

Ranius, T. (2002) *Osmoderma eremita* as an indicator of species richness of beetles in tree hollows. *Biodiversity and Conservation* 11, 931–941.

Ranius, T. (2007) Extinction risks in metapopulations of a beetle inhabiting hollow trees predicted from time series. *Ecography* 30, 716–726.

Ranius, T. & Hedin, J. (2000) The dispersal rate of a beetle, *Osmoderma eremita*, living in tree hollows. *Oecologia* 126, 363–370.

Ranius, T., Aguado, L.O., Antonsson, K. *et al.* (2005) *Osmoderma eremita* (Coleoptera, Scarabaeidae, Cetoniinae) in Europe. *Animal Biodiversity and Conservation* 28, 1–44.

Reddell, J.R. (1994) The cave faunas of Texas with special references to the Western Edwards Plateau. In: *The Caves and Karst of Texas* (eds W.R. Elliott & G. Veni), pp. 31–50. National Speleological Society, Huntsville, AL.

Ribera, I. (2000) Biogeography and conservation of Iberian water beetles. *Biological Conservation* 92, 131–150.

Ribera, I., Doledec, S., Downie, I.S. & Foster, G.N. (2001) Effect of land disturbance and stress on species traits of ground beetle assemblages. *Ecology* 82, 1112–1129.

Ribera, I., Foster, G.N. & Vogler, A.P. (2003) Does habitat use explain large scale species richness patterns of aquatic beetles in Europe? *Ecography* 26, 145–152.

Ridsdill-Smith, T.J. (1988) Survival and reproduction of *Musca vetustissima* Walker (Diptera: Muscidae) and a scarabaeine dung beetle in dung of cattle treated with avermectin B1. *Journal of the Australian Entomological Society* 27, 175–178.

Riecken, U. & Raths, U. (1996) Use of radio telemetry for studying dispersal and habitat use of *Carabus coriaceus* L. *Annales Zoologicae Fennici* 33, 109–116.

Rink, M. & Sinsch, U. (2007) Radio-telemetric monitoring of dispersing stag beetles: implications for conservation. *Journal of Zoology* 272, 235–243.

Romero-Alcaraz, E. & Avila, J.M. (2000) Landscape heterogeneity in relation to variations in epigaeic beetle diversity of a Mediterranean ecosystem. Implications for conservation. *Biodiversity and Conservation* 9, 985–1005.

Roslin, T. (2001) Large-scale spatial ecology of dung beetles. *Ecography* 24, 511–524.

Roy, H.E. & Wajnberg, E. (eds) (2008) *From Biological Control to Invasion: the Ladybird* Harmonia axyridis *as a Model Species*. Springer, Dordrecht.

Rukke, B.A. & Midtgaard, F. (1998) The importance of scale and spatial variables for the fungivorous beetle *Bolitophagus reticulatus* (Coleoptera, Tenebrionidae) in a fragmented forest landscape. *Ecography* 21, 561–572.

Sadler, J.P., Bell, D. & Fowles, A.P. (2004) The hydroecological controls and conservation value of beetles on exposed riverine sediments in England and Wales. *Biological Conservation* 118, 41–56.

Sadler, J.P., Small, E.C., Fiszpan, H., Telfer, M.G. & Niemela, J. (2006) Investigating environmental variation and landscape characteristics of an urban–rural gradient using woodland carabid assemblages. *Journal of Biogeography* 33, 1126–1138.

Sahlin, E. & Ranius, T. (2009) Habitat availability in forests and clearcuts for saproxylic beetles associated with aspen. *Biodiversity and Conservation* 18, 631–638.

Saint-German, M., Drapeau, P. & Hebert, C. (2004) Comparison of Coleoptera assemblages from a recently burned and unburned black spruce forests of northeastern North America. *Biological Conservation* 118, 583–592.

Saint-German, M., Drapeau, P. & Buddle, C.M. (2007) Host-use patterns of saproxylic phloeophagous and xylophagous Coleoptera adults and larvae along a decay gradient in standing dead black spruce and aspen. *Ecography* 30, 737–748.

Samuelson, G.A. (2003) Review of *Rhyncogonus* of the Hawaiian Islands (Coleoptera: Curculionidae). *Bishop Museum Bulletins in Entomology*, no. 11.

Samways, M.J. (2007) Implementing ecological networks for conserving insect and other biodiversity. In: *Insect Conservation Biology* (eds A.J.A. Stewart, T.R. New & O.T. Lewis), pp. 127–143. CABI Publishing, Wallingford.

Sanchez-Fernandez, D., Abellan, P., Velasco, J. & Millan, A. (2004) Selecting areas to protect the biodiversity of aquatic ecosystems in a semiarid Mediterranean region using water beetles. *Aquatic Conservation: Marine and Freshwater Ecosystems* 14, 465–479.

Sanchez-Fernandez, D., Abellan, P., Melladl, A., Velasaco, J. & Millan, A. (2006) Are water beetles good indicators of biodiversity in Mediterranean aquatic ecosystems? The case of the Segura river basin (SE Spain). *Biodiversity and Conservation* 15, 4507–4520.

Sands, D.P.A. & New, T.R. (2002) *The Action Plan for Australian Butterflies*. Environment Australia, Canberra.

Satoh, A. & Hori, M. (2005) Microhabitat selection in larvae of six species of coastal tiger beetles in Japan. *Ecological Research* 20, 143–149.

Satoh, A., Ueda, T., Enokido, Y. & Hori, M. (2003) Patterns of species assemblages and geographical distributions associated with mandible size differences in coastal tiger beetles in Japan. *Population Ecology* 45, 67–74.

Satoh, A., Sota, T., Ueda, T., Enokido, Y., Paik, J.C. & Hori, M. (2004) Evolutionary history of coastal tiger beetles in Japan based on a comparative phylogeography of four species. *Molecular Ecology* 13, 3057–3069.

Sayama, K., Makihari, H., Inoue, T. & Okochi, I. (2005) Monitoring longicorn beetles in different forest types using collision traps baited with chemical attractants. *Bulletin of the Forestry and Forest Products Research Institute* **4**, 189–199.

Scheffler, P.Y. (2005) Dung beetle (Coleoptera: Scarabaeidae) diversity and community structure across three disturbance regimes in eastern Amazonia. *Journal of Tropical Ecology* **21**, 9–19.

Schellhorn, N.A., Lane, C.P. & Olson, D.M. (2005) The co-occurrence of an introduced biological control agent (Coleoptera: *Coccinella septempunctata*) and an endangered butterfly (Lepidoptera: *Lycaeides melissa samuelis*). *Journal of Insect Conservation* **9**, 41–47.

Schmitz, H. & Bleckmann, H. (1998) The photomechanic infrared receptor for the detection of forest fires in the beetle *Melanophila acuminata* (Coleoptera: Buprestidae). *Journal of Comparative Physiology A* **182**, 647–657.

Schops, K., Wratten, S.D. & Emberson, R.M. (1999) Life cycle, behaviour and conservation of the large endemic weevil, *Hadramphus spinipennis* on the Chatham Islands, New Zealand. *New Zealand Journal of Zoology* **26**, 55–66.

Schroeder, L.M. (2007) Retention or salvage logging of standing trees killed by the spruce bark beetle *Ips typographicus*: consequences for dead wood dynamics and biodiversity. *Scandinavian Journal of Forest Research* **22**, 524–530.

Schroeder, L.M., Ranius, T., Ekbom, B & Larsson, S. (2006) Recruitment of saproxylic beetles in high stumps created for maintaining biodiversity in a boreal forest landscape. *Canadian Journal of Forest Research* **36**, 2168–2178.

Schuster, J.C., Cano, E.B. & Reyes-Castillo, P. (2003) *Proculus*, giant Latin-American passalids: revision, phylogeny and biogeography. *Acta Zoologica Mexicana* **90**, 281–306.

Scott, M.P. (1998) The ecology and behaviour of burying beetles. *Annual Review of Entomology* **43**, 595–618.

Shaw, M.R. & Hochberg, M.E. (2001) The neglect of parasitic Hymenoptera in insect conservation strategies: the British fauna as an example. *Journal of Insect Conservation* **5**, 253–263.

Shirt, D.B. (ed.) (1987) *British Red Data Books 2. Insects*. Nature Conservancy Council, Peterborough.

Siitonen, J. & Saaristo, L. (2000) Habitat requirements and conservation of *Pytho kolwensis*, a beetle species of old-growth boreal forest. *Biological Conservation* **94**, 211–220.

Sikes, D.S. & Raithel, C.J. (2002) A review of hypotheses of decline of the endangered American burying beetle (Silphidae: *Nicrophorus americanus* Olivier). *Journal of Insect Conservation* **6**, 103–113.

Simila, M., Kouki, J., Monkkonen, M. & Sippola, A.-L. (2002) Beetle species richness along the forest productivity gradient in northern Finland. *Ecography* **25**, 42–52.

Singh, P. & Moore, R.F. (eds) (1985) *Handbook of Insect Rearing*, vol. 1. Elsevier, Amsterdam.

Small, E.C., Sadler, J.P. & Telfer, M.G. (2003) Carabid beetle assemblages on urban derelict sites in Birmingham, UK. *Journal of Insect Conservation* **6**, 233–246.

Small, E., Sadler, J.P. & Telfer, M. (2006) Do landscape factors affect brownfield carabid assemblages? *Science of the Total Environment* **360**, 205–222.

Smith, M.N. (2003) *National Stag Beetle Survey 2002*. People's Trust for Endangered Species, London.

Southwood, T.R.E. (1962) Migration of terrestrial arthropods in relation to habitat. *Biological Reviews* **37**, 171–214.

Spector, S. & Ayzama, S. (2003) Rapid turnover and edge effects in dung beetle assemblages (Scarabaeidae) at a Bolivian Neotropical forest–savanna ecotone. *Biotropica* **35**, 394–404.

Speight, M.C.D. (1989) *Saproxylic Invertebrates and their Conservation*. Nature and Environment Series no. 46. Council of Europe, Strasbourg.

Spence, J.R. & Spence, D.H. (1988) Of ground-beetles and men: introduced species and synanthropic fauna of western Canada. *Memoirs of the Entomological Society of Canada* **144**, 151–168.

Spence, J.R., Langor, D.W., Niemela, J., Carcamo, H. & Currie, C.R. (1996) Northern forests and carabids: the case for concern about old-growth species. *Annales Zoologicae Fennici* **33**, 173–184.

Spence, J.R., Langor, D.W., Jacobs, J.M., Work, T.T. & Volney, W.J.A. (2008) Conservation of forest-dwelling arthropod species: simultaneous management of many small and heterogeneous risks. *Canadian Entomologist* **140**, 510–525.

Stanaway, M.A., Zalucki, M.P., Gillespie, P.S., Rodriguez, C.M. & Maynard, G.V. (2001) Pest risk assessment of insects in sea cargo containers. *Australian Journal of Entomology* **40**, 180–192.

Stork, N.E. (ed.) (1990) *The Role of Ground Beetles in Ecological and Environmental Studies*. Intercept, Andover.

Stork, N.E. (1991) The composition of the arthropod fauna of Bornean lowland rain forest trees. *Journal of Tropical Ecology* 7, 161–189.

Stork, N.E. (1997) Measuring global biodiversity and its decline. In: *Biodiversity II. Understanding and Protecting Our Biological Resources* (eds M.L. Reaka-Kudla, D.E. Wilson & E.O. Wilson), pp. 41–68. Joseph Henry Press, Washington, DC.

Strong, A.M., Dickert, C.A. & Bell, R.T. (2002) Ski trail effects on a beetle (Coleoptera: Carabidae, Elateridae) community in Vermont. *Journal of Insect Conservation* 6, 149–159.

Strong, L. (1992) Avermectins: a review of their impacts on insects of cattle dung. *Bulletin of Entomological Research* 82, 265–274.

Suominen, O., Niemela, J., Martikainen, P., Nielemla, P. & Kojola, I (2003) Impact of reindeer grazing on ground-dwelling Carabidae and Curculionidae assemblages in Lapland. *Ecography* 26, 503–513.

Svensson, G.P., Larsson, M.C. & Hedin, J. (2003) Air sampling of its pheromone to monitor the occurrence of *Osmoderma eremita*, a threatened beetle inhabiting hollow trees. *Journal of Insect Conservation* 7, 189–198.

Svensson, G.P., Larsson, M.C. & Hedin, J. (2004) Attraction of the larval predator *Elater ferrugineus* to the sex pheromone of its prey, *Osmoderma eremita*, and its implication for conservation biology. *Journal of Chemical Ecology* 30, 353–363.

Sverdrup-Thygeson, A. (2001) Can 'continuity indicator species' predict species richness or red-listed species of saproxylic beetles? *Biodiversity and Conservation* 10, 815–832.

Takeda, M., Amano, T., Katoh, K. & Higuchi, H. (2006) The habitat requirements of the Genji-firefly *Luciola cruciata* (Coleoptera: Lampyridae), a representative endemic species of Japanese rural landscapes. *Biodiversity and Conservation* 15, 191–203.

Talley, T.S. (2007) Which spatial heterogeneity framework? Consequences for conclusions about patchy population distributions. *Ecology* 8, 1476–1489.

Talley, T.S., Fleischman, E., Holyoak, M., Murphy, D. & Ballard, A. (2007) Rethinking a rare species conservation strategy in an urbanising landscape: the case of the valley elderberry longhorn beetle. *Biological Conservation* 135, 21–32.

Talley, T.S., Wright, D. & Holyoak, M. (2006) Assistance with the 5-year review of the Valley elderberry longhorn beetle (*Desmocerus californicus dimorphus*). US Fish and Wildlife Service, Sacramento.

Tansy, C.L. (2006) *Hungerford's Crawling Water Beetle* (Brychius hungerfordi) *Recovery Plan*. United States Fish and Wildlife Service, Fort Snelling, MN.

Tavakilian, G., Berkov, A., Meurer-Grimes, B. & Mori, S. (1997) Neotropical tree species and their faunas of xylophagous longicorns (Coleoptera: Cerambycidae) in French Guiana. *Botanical Review* 63, 303–355.

Taylor, R.W. (1983) Descriptive taxonomy: past, present and future. In: *Australian Systematic Entomology: a Bicentenary Perspective* (eds E. Highley & R.W. Taylor), pp. 93–134. CSIRO Publishing, Melbourne.

Telfer, M.G. & Eversham, B.C. (1996) Ecology and conservation of heathland Carabidae in eastern England. *Annales Zoologicae Fennici* 33, 133–138.

Thiele, H.-U. (1977) *Carabid Beetles in their Environments*. Springer-Verlag, Berlin and Heidelberg.

Thomaes, A., Kervyn, T. & Maes, D. (2008) Applying species distribution modeling for the conservation of the threatened saproxylic stag beetle *Lucanus cervus*. *Biological Conservation* 141, 1400–1410.

Thomas, C.F.G., Holland, J.M. & Brown, J.M. (2002) The spatial distribution of carabid beetles in agricultural landscapes. In: *The Agroecology of Carabid Beetles* (ed. J.M. Holland), pp. 305–344. Intercept, Andover.

Thomas, J.A. (1983) The ecology and conservation of *Lysandra bellargus* (Lepidoptera: Lycaenidae) in Britain. *Journal of Applied Ecology* 20, 59–83.

Thomas, M.B., Wratten, S.D. & Sotherton, N.W. (1991) Creation of 'island' habitats in farmland to manipulate populations of beneficial arthropods: predator densities and emigration. *Journal of Applied Ecology* 28, 906–917.

Thomas, M.B., Wratten, S.D. & Sotherton, N.W. (1992) Creation of 'island' habitats in farmland to manipulate populations of beneficial arthropods: predator densities and species composition. *Journal of Applied Ecology* 29, 524–531.

Thomas, S.R., Goulson, D. & Holland, J.M. (2000) The contribution of beetle banks to farmland biodiversity. *Aspects of Applied Biology* 62, 31–38.

Threatened Species Unit (2000a) *Listing Statement Ida Bay Cave Beetle* Idacarabus troglodytes. Nature Conservation Branch, Department of Primary Industry, Water and Environment, Tasmania.

Threatened Species Unit (2000b) *Listing Statement Blind Cave Beetle* Goedetrechus mendurnae. Nature Conservation Branch, Department of Primary Industry, Water and Environment, Tasmania.

Timms, B.V. & Hammer, U.T. (1988) Water beetles of some saline lakes in Saskatchewan, Canada. *Canadian Field-Naturalist* **102**, 246–250.

Toda, Y. & Sakuratani, Y. (2006) Expansion of the geographical distribution of an exotic ladybird beetle, *Adalia bipunctata* (Coleoptera; Coccinellidae), and its interspecific relationships with native ladybird beetles in Japan. *Ecological Research* **21**, 292–300.

Toivanen, T. & Koticho, J.S. (2007) Mimicking natural disturbances of boreal forests: the effects of controlled burning and creating dead wood on beetle diversity. *Biodiversity and Conservation* **16**, 3193–3211.

Topp, W., Kappes, H., Kulfan, J. & Zach, P. (2006) Litter-dwelling beetles in primeval forests of Central Europe: does dead wood matter? *Journal of Insect Conservation* **10**, 229–239.

Trumbo, S.T. & Bloch, P.L. (2000) Habitat fragmentation and burying beetle (Coleoptera: Silphidae) communities. *Journal of Insect Conservation* **4**, 2435–252.

Tshikae, B.P., Davis, A.L.V. & Scholtz, C.H. (2008) Trophic associations of a dung beetle assemblage (Scarabaeidae: Scarabaeinae) in a woodland savanna of Botswana. *Environmental Entomology* **37**, 431–441.

Tung, V.W.-Y. (1983) *Common Malaysian Beetles*. Malayan Nature Handbooks. Longman, London.

Turin, H. & den Boer, P.J. (1988) Changes in the distribution of carabid beetles in The Netherlands since 1880. II. Isolation of habitats and long-term time trends in the occurrence of carabid species with different powers of dispersal (Coleoptera, Carabidae). *Biological Conservation* **44**, 179–200.

Turner, C.R. (2007) Water beetles associated with reservoirs on Table Mountain, Cape Town: implications for conservation. *Journal of Insect Conservation* **11**, 75–83.

Tyndale-Biscoe, M. (1990) *Common Dung Beetles in Pastures of South-eastern Australia*. CSIRO, Melbourne.

United States Fish and Wildlife Service (1994) *Endangered Karst Invertebrates (Travis and Williamson Counties, Texas) Recovery Plan*. USFWS, Albuquerque, NM.

Vanbergen, A.J., Woodcock, B.A., Watt, A.D. & Niemela, J. (2005) Effect of land-use heterogeneity on carabid communities at the landscape scale. *Ecography* **28**, 3–16.

Vandermeer, J. & Perfecto, I. (1985) *Breakfast of Biodiversity: the Truth about Rainforest Defoliation*. Food First Books, Oakland, CA.

Varchola, J.M. & Dunn, J.P. (2001) Influence of hedgerow and grassy field borders on ground beetle (Coleoptera: Carabidae) activity in fields of corn. *Agriculture, Ecosystems and Environment* **83**, 153–163.

Verdu, J.R., Moreno, C.E., Sanchez-Rojas, G., Numa, C., Galante, E. & Halffter, G. (2007) Grazing promotes dung beetle diversity in the xeric landscape of a Mexican Biosphere Reserve. *Biological Conservation* **140**, 308–317.

Vernes, K., Pope, L.C., Hill, C.J. & Barlocher, F. (2005) Seasonality, dung specificity and competition in dung beetle assemblages in the Australian Wet Tropics, north-eastern Australia. *Journal of Tropical Ecology* **21**, 1–8.

Vogler, A.P. & DeSalle, R. (1994) Diagnosing units of conservation management. *Conservation Biology* **8**, 354–363.

Wagner, T. (2003) Seasonality of canopy beetles in Uganda. In: *Arthropods of Tropical Forests: Spatiotemporal Dynamics and Resource Use in the Canopy* (eds Y. Basset, R. Kitching, S. Miller & V. Novotny), pp. 146–158. Cambridge University Press, Cambridge.

Walsh, G.B. & Dibb, J.R. (eds) (1954) *A Coleopterist's Handbook*. Amateur Entomologists Society, London.

Warren, M.S., Hill, J.K., Thomas, J.A. *et al.* (2001) Rapid responses of British butterflies to opposing forces of climate and habitat change. *Nature* **414**, 65–69.

Watt, J.C. (1979) Conservation of the Cromwell Chafer *Prodontria lewisi* (Coleoptera: Scarabaeidae). *New Zealand Journal of Ecology* **2**, 22–29.

Watts, C.H. & Lariviere, M.-C. (2004) The importance of urban reserves for conserving beetle communities: a case study from New Zealand. *Journal of Insect Conservation* **8**, 47–58.

Webber, J.F. & Gibbs, J.N. (1989) Insect dissemination of fungal pathogens of trees. In: *Insect–Fungus Interactions* (eds N. Wilding, N.M. Collins, P.M. Hammond & J.F. Webber), pp 161–193. Academic Press, London.

Weber, F. & Heimbach, U. (2001) Behavioural, reproductive and developmental seasonality in *Carabus auronitens* and *Carabus nemoralis* (Col., Carabidae). A demographic comparison between two co-existing spring breeding populations and tests for intra- and interspecific competition and for synchronizing weather events. *Mitteilungen aus den Biologischen Bundesanstadt fur Land- und Forstschaft, Berlin-Dahlem*, Heft 382.

Weir, J.S. (1972) Diversity and abundance of aquatic insects reduced by introduction of the fish *Clarias gariepinus* to pools in Central Africa. *Biological Conservation* 4, 169–174.

Welch, R.C. (1990) Dispersal of invertebrates in the agricultural environment. In: *Species Dispersal in Agricultural Habitats* (eds R.G.H. Bunce & D.C. Howard), pp. 203–218. Belhaven Press, London.

Wells, S.M., Pyle, R.M. & Collins, N.M. (1983) *IUCN Invertebrate Red Data Book*. IUCN, Gland, Switzerland.

Whitehouse, N.J. (2006) The Holocene British and Irish ancient forest fossil beetle fauna: implications for forest history, biodiversity, and faunal colonization. *Quaternary Science Reviews* 25, 1755–1789.

Whitford, W.G. (2000) Keystone arthropods as webmasters in desert ecosystems. In: *Invertebrates as Webmasters in Ecosystems* (eds D.C. Coleman & P.F. Hendrix), pp. 23–41. CABI Publishing, Wallingford.

Wikars, L.-O. (2002) Dependence on fire in wood-living insects: an experiment with burned and unburned spruce and birch logs. *Journal of Insect Conservation* 6, 1–12.

Wilson, D.S., Knollenberg, W.G. & Fudge, J. (1984) Species packing and temperature dependent competition among burying beetles (Silphidae: *Nicrophorus*). *Ecological Entomology* 9, 205–216.

Wilson, R.J., Davies, Z.G. & Thomas, C.D. (2007) Insects and climate change: processes, patterns and implications for conservation. In: *Insect Conservation Biology* (eds A.J.A. Stewart, T.R. New & O.T. Lewis), pp. 245–279. CABI Publishing, Wallingford.

Wolda, H. (1988) Insect seasonality: why? *Annual Review of Ecology and Systematics* 19, 1–19.

Wolda, H., O'Brien, C.W. & Stockwell, H.P. (1998) Weevil diversity and seasonality in tropical Panama as deduced from light-trap catches (Coleoptera: Curculionoidea). *Smithsonian Contributions to Zoology*, no. 590.

Work, T.T., Koivula, M., Klimaszewski, J. *et al.* (2008) Evaluation of carabid beetles as indicators of forest change in Canada. *Canadian Entomologist* 140, 393–414.

Wratten, S.D. & Thomas, C.F.G. (1990) Farm-scale spatial dynamics of predators and parasitoids in agricultural landscapes. In: *Species Dispersal in Agricultural Habitats* (eds R.G.H. Bunce & D.C. Howard), pp. 219–237. Belhaven Press, London.

Yeates, D.K., Bouchard, P. & Monteith, G.B. (2002) Patterns and levels of endemism in the Australian Wet Tropics rainforest: evidence from flightless insects. *Invertebrate Systematics* 16, 605–619.

Yeates, D.K., Harvey, M.S. & Austin, A.D. (2003) New estimates for terrestrial arthropod species-richness in Australia. *Records of the South Australian Museum, Monograph Series*, no. 7, 231–241.

Yuma, M. (2007) Effect of rainfall on the long-term population dynamics of the aquatic firefly *Luciola cruciata*. *Entomological Science* 10, 237–244.

Zahn, A., Juen, A., Traugott, M. & Lang, A. (2007) Low density cattle grazing enhances arthropod diversity of abandoned wetland. *Applied Ecology and Environmental Research* 5, 73–86.

Zeh, D.W., Zeh, J.A. & Bonilla, M.M. (2003) Phylogeography of the giant harlequin beetle (*Acrocinus longimanus*). *Journal of Biogeography* 30, 747–753.

Zimmerman, E.C. (1994) *Australian Weevils. Volume I. Anthribidae to Attelabidae*. CSIRO Publishing, Melbourne.

Zwoelfer, H. & Harris, P. (1984) Biology and host specificity of *Rhinocyllus conicus* (Froel.) (Col., Curculionidae), a successful agent for biocontrol of the thistle, *Carduus nutans* L. *Zeitschrift fur Angewandte Entomologie* 97, 36–62.

Index